[illegible]

Titu Andre[illegible] [illegible] Thomson, K[illegible]
[illegible] in Harmonic Analysis and
Applicat[illegible] [illegible] at the University of Texas
[illegible] of the USA Mathematical Olympiad, served as di-
[illegible] MAA American Mathematics Competitions (1998–2003), coach of
[illegible] and Mathematical Olympiad Team (IMO) for 10 years (1993–
2002) [illegible] the Mathematical Olympiad Summer Program (1995–2002),
[illegible] USA IMO Team (1995–2002). [illegible]
[illegible] Board, [illegible] prestigious
[illegible]. The [illegible] his directorship
the [illegible] Math [illegible] the Edyth May Sliffe
Award for Distinguished High School Mathematics Teaching [illegible]
1994 [illegible] MAA in 2005
[illegible] Mathematical Olympiad [illegible]
[illegible] in [illegible] at the
[illegible] and problem books
[illegible]

Dorin Andrica [illegible] 1992 [illegible]
[illegible] University
[illegible] has been [illegible]
the Department of [illegible]
[illegible]
[illegible]
[illegible]
[illegible]
[illegible]
[illegible] journals [illegible]
[illegible]
[illegible] USA Mathematical [illegible]
[illegible] and [illegible] Summer Program [illegible]

Zuming Feng [illegible] University with [illegible]
[illegible] at Phillips [illegible]
[illegible] was
[illegible] 2002, [illegible]
[illegible] mathematical [illegible] (1999–2002) [illegible]
[illegible] since 1998 [illegible]
[illegible] USA Mathe-
matical Olympiad [illegible]
[illegible] (MOSP) since 2003.
He received the Edyth May Sliffe Award for Distinguished High School Mathematics Teaching from the MAA in 1996 and 2002.

About the Authors

Titu Andreescu received his Ph.D. from the West University of Timisoara, Romania. The topic of his dissertation was "Research on Diophantine Analysis and Applications." Professor Andreescu currently teaches at The University of Texas at Dallas. He is past chairman of the USA Mathematical Olympiad, served as director of the MAA American Mathematics Competitions (1998–2003), coach of the USA International Mathematical Olympiad Team (IMO) for 10 years (1993–2002), director of the Mathematical Olympiad Summer Program (1995–2002), and leader of the USA IMO Team (1995–2002). In 2002 Titu was elected member of the IMO Advisory Board, the governing body of the world's most prestigious mathematics competition. Titu co-founded in 2006 and continues as director of the AwesomeMath Summer Program (AMSP). He received the Edyth May Sliffe Award for Distinguished High School Mathematics Teaching from the MAA in 1994 and a "Certificate of Appreciation" from the president of the MAA in 1995 for his outstanding service as coach of the Mathematical Olympiad Summer Program in preparing the US team for its perfect performance in Hong Kong at the 1994 IMO. Titu's contributions to numerous textbooks and problem books are recognized worldwide.

Dorin Andrica received his Ph.D in 1992 from "Babeş-Bolyai" University in Cluj-Napoca, Romania; his thesis treated critical points and applications to the geometry of differentiable submanifolds. Professor Andrica has been chairman of the Department of Geometry at "Babeş-Bolyai" since 1995. He has written and contributed to numerous mathematics textbooks, problem books, articles and scientific papers at various levels. He is an invited lecturer at university conferences around the world: Austria, Bulgaria, Czech Republic, Egypt, France, Germany, Greece, Italy, the Netherlands, Portugal, Serbia, Turkey, and the USA. Dorin is a member of the Romanian Committee for the Mathematics Olympiad and is a member on the editorial boards of several international journals. Also, he is well known for his conjecture about consecutive primes called "Andrica's Conjecture." He has been a regular faculty member at the Canada–USA Mathcamps between 2001–2005 and at the AwesomeMath Summer Program (AMSP) since 2006.

Zuming Feng received his Ph.D. from Johns Hopkins University with emphasis on Algebraic Number Theory and Elliptic Curves. He teaches at Phillips Exeter Academy. Zuming also served as a coach of the USA IMO team (1997–2006), was the deputy leader of the USA IMO Team (2000–2002), and an assistant director of the USA Mathematical Olympiad Summer Program (1999–2002). He has been a member of the USA Mathematical Olympiad Committee since 1999, and has been the leader of the USA IMO team and the academic director of the USA Mathematical Olympiad Summer Program since 2003. Zuming is also co-founder and academic director of the AwesomeMath Summer Program (AMSP) since 2006. He received the Edyth May Sliffe Award for Distinguished High School Mathematics Teaching from the MAA in 1996 and 2002.

Titu Andreescu
Dorin Andrica
Zuming Feng

104 Number Theory Problems

From the Training of the USA IMO Team

Birkhäuser
Boston • Basel • Berlin

Titu Andreescu
The University of Texas at Dallas
Department of Science/Mathematics Education
Richardson, TX 75083
U.S.A.
titu.andreescu@utdallas.edu

Dorin Andrica
"Babeş-Bolyai" University
Faculty of Mathematics
3400 Cluj-Napoca
Romania
dorinandrica@yahoo.com

Zuming Feng
Phillips Exeter Academy
Department of Mathematics
Exeter, NH 03833
U.S.A.
zfeng@exeter.edu

Cover design by Mary Burgess.

Mathematics Subject Classification (2000): 00A05, 00A07, 11-00, 11-XX, 11Axx, 11Bxx, 11D04

Library of Congress Control Number: 2006935812

ISBN-10: 0-8176-4527-6 e-ISBN-10: 0-8176-4561-6
ISBN-13: 978-0-8176-4527-4 e-ISBN-13: 978-0-8176-4561-8

Printed on acid-free paper.

9 8 7 6 5 4 3 2 1

www.birkhauser.com (EB)

Contents

Preface vii

Acknowledgments ix

Abbreviations and Notation xi

1 Foundations of Number Theory 1
- Divisibility 1
- Division Algorithm 4
- Primes 5
- The Fundamental Theorem of Arithmetic 7
- G.C.D. 11
- Euclidean Algorithm 12
- Bézout's Identity 13
- L.C.M. 16
- The Number of Divisors 17
- The Sum of Divisors 18
- Modular Arithmetics 19
- Residue Classes 24
- Fermat's Little Theorem and Euler's Theorem 27
- Euler's Totient Function 33
- Multiplicative Function 36
- Linear Diophantine Equations 38
- Numerical Systems 40
- Divisibility Criteria in the Decimal System 46
- Floor Function 52
- Legendre's Function 65
- Fermat Numbers 70
- Mersenne Numbers 71
- Perfect Numbers 72

2 Introductory Problems 75

3 Advanced Problems 83

4 Solutions to Introductory Problems 91

5 Solutions to Advanced Problems 131

Glossary 189

Further Reading 197

Index 203

Preface

This book contains 104 of the best problems used in the training and testing of the U.S. International Mathematical Olympiad (IMO) team. It is not a collection of very difficult, and impenetrable questions. Rather, the book gradually builds students' number-theoretic skills and techniques. The first chapter provides a comprehensive introduction to number theory and its mathematical structures. This chapter can serve as a textbook for a short course in number theory. This work aims to broaden students' view of mathematics and better prepare them for possible participation in various mathematical competitions. It provides in-depth enrichment in important areas of number theory by reorganizing and enhancing students' problem-solving tactics and strategies. The book further stimulates students' interest for the future study of mathematics.

In the United States of America, the selection process leading to participation in the International Mathematical Olympiad (IMO) consists of a series of national contests called the American Mathematics Contest 10 (AMC 10), the American Mathematics Contest 12 (AMC 12), the American Invitational Mathematics Examination (AIME), and the United States of America Mathematical Olympiad (USAMO). Participation in the AIME and the USAMO is by invitation only, based on performance in the preceding exams of the sequence. The Mathematical Olympiad Summer Program (MOSP) is a four-week intensive training program for approximately fifty very promising students who have risen to the top in the American Mathematics Competitions. The six students representing the United States of America in the IMO are selected on the basis of their USAMO scores and further testing that takes place during MOSP. Throughout MOSP, full days of classes and extensive problem sets give students thorough preparation in several important areas of mathematics. These topics include combinatorial arguments and identities, generating functions, graph theory, recursive relations, sums and products, probability, number theory, polynomials, functional equations, complex numbers in geometry, algorithmic proofs, combinatorial and advanced geometry, functional equations, and classical inequalities.

Olympiad-style exams consist of several challenging essay problems. Correct solutions often require deep analysis and careful argument. Olympiad questions can seem impenetrable to the novice, yet most can be solved with elementary high school mathematics techniques, when cleverly applied.

Here is some advice for students who attempt the problems that follow.

- Take your time! Very few contestants can solve all the given problems.

- Try to make connections between problems. An important theme of this work is that all important techniques and ideas featured in the book appear more than once!

- Olympiad problems don't "crack" immediately. Be patient. Try different approaches. Experiment with simple cases. In some cases, working backward from the desired result is helpful.

- Even if you can solve a problem, do read the solutions. They may contain some ideas that did not occur in your solutions, and they may discuss strategic and tactical approaches that can be used elsewhere. The solutions are also models of elegant presentation that you should emulate, but they often obscure the tortuous process of investigation, false starts, inspiration, and attention to detail that led to them. When you read the solutions, try to reconstruct the thinking that went into them. Ask yourself, "What were the key ideas? How can I apply these ideas further?"

- Go back to the original problem later, and see whether you can solve it in a different way. Many of the problems have multiple solutions, but not all are outlined here.

- Meaningful problem solving takes practice. Don't get discouraged if you have trouble at first. For additional practice, use the books on the reading list.

Titu Andreescu
Dorin Andrica
Zuming Feng

October 2006

Acknowledgments

Thanks to Sara Campbell, Yingyu (Dan) Gao, Sherry Gong, Koene Hon, Ryan Ko, Kevin Medzelewski, Garry Ri, and Kijun (Larry) Seo. They were the members of Zuming's number theory class at Phillips Exeter Academy. They worked on the first draft of the book. They helped proofread the original manuscript, raised critical questions, and provided acute mathematical ideas. Their contribution improved the flavor and the structure of this book. We thank Gabriel Dospinescu (Dospi) for many remarks and corrections to the first draft of the book. Some materials are adapted from [11], [12], [13], and [14]. We also thank those students who helped Titu and Zuming edit those books.

Many problems are either inspired by or adapted from mathematical contests in different countries and from the following journals:

- *The American Mathematical Monthly*, United States of America
- *Crux*, Canada
- *High School Mathematics*, China
- *Mathematics Magazine*, United States of America
- *Revista Matematică Timişoara*, Romania

We did our best to cite all the original sources of the problems in the solution section. We express our deepest appreciation to the original proposers of the problems.

Abbreviations and Notation

Abbreviations

AHSME	American High School Mathematics Examination
AIME	American Invitational Mathematics Examination
AMC10	American Mathematics Contest 10
AMC12	American Mathematics Contest 12, which replaces AHSME
APMC	Austrian–Polish Mathematics Competition
ARML	American Regional Mathematics League
Balkan	Balkan Mathematical Olympiad
Baltic	Baltic Way Mathematical Team Contest
HMMT	Harvard–MIT Math Tournament
IMO	International Mathematical Olympiad
USAMO	United States of America Mathematical Olympiad
MOSP	Mathematical Olympiad Summer Program
Putnam	The William Lowell Putnam Mathematical Competition
St. Petersburg	St. Petersburg (Leningrad) Mathematical Olympiad

Notation for Numerical Sets and Fields

$\mathbb{Z}$	the set of integers
$\mathbb{Z}_n$	the set of integers modulo n
$\mathbb{N}$	the set of positive integers
$\mathbb{N}_0$	the set of nonnegative integers
$\mathbb{Q}$	the set of rational numbers
$\mathbb{Q}^+$	the set of positive rational numbers
$\mathbb{Q}^0$	the set of nonnegative rational numbers
$\mathbb{Q}^n$	the set of n-tuples of rational numbers
$\mathbb{R}$	the set of real numbers
$\mathbb{R}^+$	the set of positive real numbers
$\mathbb{R}^0$	the set of nonnegative real numbers
$\mathbb{R}^n$	the set of n-tuples of real numbers
$\mathbb{C}$	the set of complex numbers
$[x^n](p(x))$	the coefficient of the term x^n in the polynomial $p(x)$

Notation for Sets, Logic, and Number Theory

$\lvert A \rvert$	the number of elements in the set A
$A \subset B$	A is a proper subset of B
$A \subseteq B$	A is a subset of B
$A \setminus B$	A without B (set difference)
$A \cap B$	the intersection of sets A and B
$A \cup B$	the union of sets A and B
$a \in A$	the element a belongs to the set A
$n \mid m$	n divides m
$\gcd(m, n)$	the greatest common divisor of m, n
$\operatorname{lcm}(m, n)$	the least common multiple of m, n
$\pi(n)$	the number of primes $\leq n$
$\tau(n)$	number of divisors of n
$\sigma(n)$	sum of positive divisors of n
$a \equiv b \pmod{m}$	a and b are congruent modulo m
φ	Euler's totient function
$\operatorname{ord}_m(a)$	order of a modulo m
μ	Möbius function
$\overline{a_k a_{k-1} \ldots a_0}_{(b)}$	base-b representation
$S(n)$	the sum of digits of n
$(f_1, f_2, \ldots, f_m)$	factorial base expansion
$\lfloor x \rfloor$	floor of x
$\lceil x \rceil$	ceiling of x
$\{x\}$	fractional part of x
e_p	Legendre's function
$p^k \Vert n$	p^k fully divides n
f_n	Fermat number
M_n	Mersenne number

1

Foundations of Number Theory

Divisibility

Back in elementary school, we learned four fundamental operations on numbers (integers), namely, addition (+), subtraction (−), multiplication (× or ·), and division (÷ or / or $\frac{-}{c}$). For any two integers a and b, their sum $a+b$, differences $a-b$ and $b-a$, and product ab are all integers, while their quotients $a \div b$ (or a/b or $\frac{a}{b}$) and $b \div a$ are not necessarily integers.

For an integer m and a nonzero integer n, we say that m **is divisible by** n or n **divides** m if there is an integer k such that $m = kn$; that is, $\frac{m}{n}$ is an integer. We denote this by $n \mid m$. If m is divisible by n, then m is called a **multiple** of n; and n is called a **divisor** (or **factor**) of m.

Because $0 = 0 \cdot n$, it follows that $n \mid 0$ for all integers n. For a fixed integer n, the multiples of n are $0, \pm n, \pm, 2n, \ldots$. Hence it is not difficult to see that there is a multiple of n among every n consecutive integers. If m is not divisible by n, then we write $n \nmid m$. (Note that $0 \nmid m$ for all nonzero integers m, since $m \neq 0 = k \cdot 0$ for all integers k.)

Proposition 1.1. Let x, y, and z be integers. We have the following basic properties:

(a) $x \mid x$ (reflexivity property);

(b) If $x \mid y$ and $y \mid z$, then $x \mid z$ (transitivity property);

(c) If $x \mid y$ and $y \neq 0$, then $|x| \leq |y|$;

(d) If $x \mid y$ and $x \mid z$, then $x \mid \alpha y + \beta z$ for any integers α and β;

(e) If $x \mid y$ and $x \mid y \pm z$, then $x \mid z$;

(f) If $x \mid y$ and $y \mid x$, then $|x| = |y|$;

(g) If $x \mid y$ and $y \neq 0$, then $\frac{y}{x} \mid y$;

(h) for $z \neq 0$, $x \mid y$ if and only if $xz \mid yz$.

The proofs of the above properties are rather straightforward from the definition. We present these proofs only to give the reader some relevant examples of writing proofs.

Proof: For (a), we note that $x = 1 \cdot x$. In (b) to (h), the condition $x \mid y$ is given; that is, $y = kx$ for some integer k.

For (b), we have $y \mid z$; that is, $z = k_1 y$ for some integer k_1. Then $z = (kk_1)x$, or $x \mid z$.

For (c), we note that if $y \neq 0$, then $|k| \geq 1$, and so $|y| = |k| \cdot |x| \geq |x|$.

For (d), we further assume that $z = k_2 x$. Then $\alpha y + \beta z = (k\alpha + k_2\beta)x$.

For (e), we obtain $y \pm z = k_3 x$, or $\pm z = k_3 x - y = (k_3 - k)x$. It follows that $z = \pm(k - k_3)x$.

For (f), because $x \mid y$ and $y \mid x$, it follows that $x \neq 0$ and $y \neq 0$. By (c), we have $|y| \geq |x|$ and $|x| \geq |y|$. Hence $|x| = |y|$.

For (g), $\frac{y}{x} = k \neq 0$ is an integer. Since $y = x \cdot k$, $k \mid y$.

For (h), since $z \neq 0$, $x \neq 0$ if and only if $xz \neq 0$. Note that $y = kx$ if and only if $yz = kxz$. □

The property (g) is simple but rather helpful. For a nonzero integer n, there is an even number of positive divisors of n unless n is a **perfect square**; that is, $n = m^2$ for some integer m. (If an integer is not divisible by any perfect square, then it is called **square free**. If $n = m^3$ for some integer m, then n is called a **perfect cube**. In general, if $n = m^s$ for integers m and s with $s \geq 2$, then n is called a **perfect power**.) This is because all the divisors of y appear in pairs, namely, x and $\frac{y}{x}$ (observe that $x \neq \frac{y}{x}$ if y is not a perfect square). Here is a classic brain teaser:

Example 1.1. Twenty bored students take turns walking down a hall that contains a row of closed lockers, numbered 1 to 20. The first student opens all the lockers; the second student closes all the lockers numbered 2, 4, 6, 8, 10, 12, 14, 16, 18, 20; the third student operates on the lockers numbered 3, 6, 9, 12, 15, 18: if a locker was closed, he opens it, and if a locker was open, he closes it; and so on. For the ith student, he works on the lockers numbered by multiples of i: if a locker was closed, he opens it, and if a locker was open, he closes it. What is the number of the lockers that remain open after all the students finish their walks?

Solution: Note that the ith locker will be operated by student j if and only if $j \mid i$. By property (g), this can happen if and only if the locker will also be operated by student $\frac{i}{j}$. Thus, only the lockers numbered $1 = 1^2, 4 = 2^2, 9 = 3^2,$

and $16 = 4^2$ will be operated on an odd number of times, and these are the lockers that will be left open after all the operations. Hence the answer is 4. □

The set of integers, denoted by $\mathbb{Z}$, can be partitioned into two subsets, the set of odd integers and the set of even integers:

$$\{\pm 1, \pm 3, \pm 5, \dots\} \quad \text{and} \quad \{0, \pm 2, \pm 4, \dots\},$$

respectively. Although the concepts of odd and even integers appear straightforward, they come handy in tackling various number-theoretic problems. Here are some basic ideas:

(1) an odd number is of the form $2k + 1$, for some integer k;

(2) an even number is of the form $2m$, for some integer m;

(3) the sum of two odd numbers is an even number;

(4) the sum of two even numbers is an even number;

(5) the sum of an odd and even number is an odd number;

(6) the product of two odd numbers is an odd number;

(7) a product of integers is even if and only if at least one of its factors is even.

Example 1.2. Let n be an integer greater than 1. Prove that

(a) 2^n is the sum of two odd consecutive integers;

(b) 3^n is the sum of three consecutive integers.

Proof: For (a), the relation $2^n = (2k - 1) + (2k + 1)$ implies $k = 2^{n-2}$ and we obtain $2^n = (2^{n-1} - 1) + (2^{n-1} + 1)$.

For (b), the relation $3^n = (s - 1) + s + (s + 1)$ implies $s = 3^{n-1}$ and we obtain the representation $3^n = (3^{n-1} - 1) + 3^{n-1} + (3^{n-1} + 1)$. □

Example 1.3. Let k be an even number. Is it possible to write 1 as the sum of the reciprocals of k odd integers?

Solution: The answer is negative.

We approach indirectly. Assume that

$$1 = \frac{1}{n_1} + \cdots + \frac{1}{n_k}$$

for some odd integers $n_1, \dots, n_k$; then clearing denominators we obtain

$n_1 \cdots n_k = s_1 + \cdots + s_k$, where s_i are all odd. But this is impossible since the left-hand side is odd and the right-hand side is even. □

If k is odd, such representations are possible. Here is one example for $k = 9$ and $n_1, \ldots, n_9$ are distinct odd positive integers:

$$1 = \frac{1}{3} + \frac{1}{5} + \frac{1}{7} + \frac{1}{9} + \frac{1}{11} + \frac{1}{15} + \frac{1}{35} + \frac{1}{45} + \frac{1}{231}.$$

Example 1.4. [HMMT 2004] Zach has chosen five numbers from the set $\{1, 2, 3, 4, 5, 6, 7\}$. If he told Claudia what the product of the chosen numbers was, that would not be enough information for Claudia to figure out whether the sum of the chosen numbers was even or odd. What is the product of the chosen numbers?

Solution: The answer is 420.

Providing the product of the chosen numbers is equivalent to telling the product of the two unchosen numbers. The only possible products that are achieved by more than one pair of numbers are 12 ($\{3, 4\}$ and $\{2, 6\}$) and 6 ($\{1, 6\}$ and $\{2, 3\}$). But in the second case, the sum of the two (unchosen) numbers is odd (and so the five chosen numbers have odd sum too). Therefore, the first must hold, and the product of the five chosen numbers is equal to

$$\frac{1 \cdot 2 \cdot 3 \cdots 7}{12} = 420.$$ □

Division Algorithm

The following result is called the division algorithm, and it plays an important role in number theory:

Theorem 1.2a. For any positive integers a and b there exists a unique pair (q, r) of nonnegative integers such that $b = aq + r$ and $r < a$. We say that q is the **quotient** and r the **remainder** when b is divided by a.

To prove this result, we need to consider two parts: the existence of such a pair and its uniqueness.

Proof: To show the existence, we consider three cases.

(1) In this case, we assume that $a > b$. We can set $q = 0$ and $r = b < a$; that is, $(q, r) = (0, b)$.

(2) Suppose that $a = b$. We can set $q = 1$ and $r = 0 < a$; that is, $(q, r) = (1, 0)$.

(3) Finally, assume that $a < b$. There exist positive integers n such that $na > b$. Let q be the least positive integer for which $(q + 1)a > b$. Then $qa \leq b$. Let $r = b - aq$. It follows that $b = aq + r$ and $0 \leq r < a$.

Combining the three cases, we have established the existence.

For uniqueness, assume that $b = aq' + r'$, where q' and r' are also nonnegative integers satisfying $0 \le r' < a$. Then $aq + r = aq' + r'$, implying $a(q - q') = r' - r$, and so $a \mid r' - r$. Hence $|r' - r| \ge a$ or $|r' - r| = 0$. Because $0 \le r$, $r' < a$ yields $|r' - r| < a$, we are left with $|r' - r| = 0$, implying $r' = r$, and consequently, $q' = q$. □

Example 1.5. Let n be a positive integer. Prove that $3^{2^n} + 1$ is divisible by 2, but not by 4.

Proof: Clearly, 3^{2^n} is odd and $3^{2^n} + 1$ is even. Note that $3^{2^n} = (3^2)^{2^{n-1}} = 9^{2^{n-1}} = (8+1)^{2^{n-1}}$. Recall the **Binomial theorem**

$$(x+y)^m = x^m + \binom{m}{1}x^{m-1}y + \binom{m}{2}x^{m-2}y^2 + \cdots + \binom{n}{n-1}xy^{m-1} + y^m.$$

Setting $x = 8$, $y = 1$, and $m = 2^{n-1}$ in the above equation, we see that each summand besides the last (that is, $y^m = 1$) is a multiple of 8 (which is a multiple of 4). Hence the remainder of 3^{2^n} on dividing by 4 is equal to 1, and the remainder of $3^{2^n} + 1$ on dividing by 4 is equal to 2. □

The above argument can be simplified in the notation of congruence modulo 4. Congruence is an important part of number theory. We will discuss it extensively.

The division algorithm can be extended for integers:

Theorem 1.2b. For any integers a and b, $a \ne 0$, there exists a unique pair (q, r) of integers such that $b = aq + r$ and $0 \le r < |a|$.

We leave the proof of this extended version to the reader.

Primes

The integer $p > 1$ is called a **prime** (or a **prime number**) if there is no integer d with $d > 1$ and $d \ne p$ such that $d \mid p$. Any integer $n > 1$ has at least one prime divisor. If n is a prime, then that prime divisor is n itself. If n is not a prime, then let $a > 1$ be its least divisor. Then $n = ab$, where $1 < a \le b$. If a were not a prime, then $a = a_1 a_2$ with $1 < a_1 \le a_2 < a$ and $a_1 \mid n$, contradicting the minimality of a.

An integer $n > 1$ that is not a prime is called **composite**. If n is a composite integer, then it has a prime divisor p not exceeding $\sqrt{n}$. Indeed, as above, $n = ab$, where $1 < a \le b$ and a is the least divisor of n. Then $n \ge a^2$; hence $a \le \sqrt{n}$. This idea belongs to the ancient Greek mathematician Eratosthenes (250 BCE).

Note that all positive **even numbers** greater than 2 are composite. In other words, 2 is the only even (and the smallest) prime. All other primes are **odd**; that is, they are not divisible by 2. The first few primes are 2, 3, 5, 7, 11, 13, 17, 19, 23, 29. How many primes are there? Are we really sure that there are infinitely many primes? Please see Theorem 1.3 below. A comparison between the number of elements in two infinite sets might be vague, but it is *obvious* that there are *more* (in the sense of density) composite numbers than primes. We see that 2 and 3 are the only consecutive primes. Odd consecutive primes such as 3 and 5, 5 and 7, 41 and 43, are called **twin primes**. It is still an open question whether there are infinitely many twin primes. Brun has shown that even if there are infinitely many twin primes, the sum of their inverses converges. The proof is however extremely difficult.

Example 1.6. Find all positive integers n for which $3n-4$, $4n-5$, and $5n-3$ are all prime numbers.

Solution: The sum of the three numbers is an even number, so at least one of them is even. The only even prime number is 2. Only $3n-4$ and $5n-3$ can be even. Solving the equations $3n-4=2$ and $5n-3=2$ yields $n=2$ and $n=1$, respectively. It is trivial to check that $n=2$ does make all three given numbers prime. □

Example 1.7. [AHSME 1976] If p and q are primes and $x^2-px+q=0$ has distinct positive integral roots, find p and q.

Solution: Let x_1 and x_2, with $x_1 < x_2$, be the two distinct positive integer roots. Then $x^2-px+q=(x-x_1)(x-x_2)$, implying that $p=x_1+x_2$ and $q=x_1x_2$. Since q is prime, $x_1=1$. Thus, $q=x_2$ and $p=x_2+1$ are two consecutive primes; that is, $q=2$ and $p=3$. □

Example 1.8. Find 20 consecutive composite numbers.

Solution: Numbers $20!+2, 20!+3, \ldots, 20!+21$ will do the trick. □

The following result by Euclid has been known for more than 2000 years:

Theorem 1.3a. There are infinitely many primes.

Proof: Assume by way of contradiction that there are only a finite number of primes: $p_1 < p_2 < \cdots < p_m$. Consider the number $P = p_1p_2\cdots p_m + 1$.

If P is a prime, then $P > p_m$, contradicting the maximality of p_m. Hence P is composite, and consequently, it has a prime divisor $p > 1$, which is one of the primes $p_1, p_2, \ldots, p_m$, say p_k. It follows that p_k divides $p_1\cdots p_k\cdots p_m + 1$. This, together with the fact that p_k divides $p_1\cdots p_k\cdots p_m$, implies p_k divides 1, a contradiction. □

Even though there are infinitely many primes, there are no particular formulas to find them. Theorem 1.3b in the next section will reveal part of the reasoning.

The Fundamental Theorem of Arithmetic

The fundamental result in arithmetic (i.e., number theory) pertains to the prime factorization of integers:

Theorem 1.4. [The Fundamental Theorem of Arithmetic] Any integer n greater than 1 has a unique representation (up to a permutation) as a product of primes.

Proof: The existence of such a representation can be obtained as follows: Let p_1 be a prime divisor of n. If $p_1 = n$, then $n = p_1$ is a prime factorization of n. If $p_1 < n$, then $n = p_1 r_1$, where $r_1 > 1$. If r_1 is a prime, then $n = p_1 p_2$, where $p_2 = r_1$ is the desired factorization of n. If r_1 is composite, then $r_1 = p_2 r_2$, where p_2 is a prime, $r_2 > 1$, and so $n = p_1 p_2 r_2$. If r_2 is a prime, then $n = p_1 p_2 p_3$, where $r_2 = p_3$ and $r_3 = 1$, and we are done. If r_2 is composite, then we continue this algorithm, obtaining a sequence of integers $r_1 > r_2 > \cdots \geq 1$. After a finite number of steps, we reach $r_{k+1} = 1$, that is, $n = p_1 p_2 \cdots p_k$.

For uniqueness, let us assume that there is at least one positive integer n that has two distinct prime factorizations; that is,

$$n = p_1 p_2 \cdots p_k = q_1 q_2 \cdots q_h$$

where $p_1, p_2, \ldots, p_k, q_1, q_2, \ldots, q_h$ are primes with $p_1 \leq p_2 \leq \cdots p_k$ and $q_1 \leq q_2 \cdots q_h$ such that the k-tuple $(p_1, p_2, \ldots, p_k)$ is not the same as the h-tuple $(q_1, q_2, \ldots, q_h)$. It is clear that $k \geq 2$ and $h \geq 2$. Let n be the *minimal such integer*. We will derive a contradiction by finding a smaller positive integer that also has two distinct prime factorizations.

We claim that $p_i \neq q_j$ for any $i = 1, 2, \ldots, k$, $j = 1, 2, \ldots, h$. If, for example, $p_k = q_h = p$, then $n' = n/p = p_1 \cdots p_{k-1} = q_1 \cdots q_{h-1}$ and $1 < n' < n$, contradicting the minimality of n. Assume without loss of generality that $p_1 \leq q_1$; that is, p_1 is the least prime factor of n in the above representations. By applying the division algorithm it follows that

$$\begin{aligned} q_1 &= p_1 c_1 + r_1, \\ q_2 &= p_1 c_2 + r_2, \\ &\vdots \\ q_h &= p_1 c_h + r_h, \end{aligned}$$

where $1 \leq r_i < p_1, i = 1, \ldots, h$.

We have

$$n = q_1 q_2 \cdots q_h = (p_1 c_1 + r_1)(p_1 c_2 + r_2) \cdots (p_1 c_h + r_h).$$

Expanding the last product we obtain $n = m p_1 + r_1 r_2 \cdots r_h$ for some positive integer m. Setting $n' = r_1 r_2 \cdots r_h$ we have $n = p_1 p_2 \cdots p_k = m p_1 + n'$. It follows that $p_1 \mid n'$ and $n' = p_1 s$. As we have shown, s can be written as a product of primes. We write $s = s_1 s_2 \cdots s_i$, where $s_1, s_2, \ldots, s_i$ are primes.

On the other hand, using the factorization of $r_1, r_2, \ldots, r_h$ into primes, all their factors are less than $r_i < p_1$. From $n' = r_1 r_2 \cdots r_h$, it follows that n' has a factorization into primes of the form $n' = t_1 t_2 \cdots t_j$, where $t_s < p_1$, $s = 1, 2, \ldots, j$. This factorization is different from $n' = p_1 s_1 s_2 \cdots s_i$. But $n' < n$, contradicting the minimality of n. □

From the above theorem it follows that any integer $n > 1$ can be written uniquely in the form

$$n = p_1^{\alpha_1} \cdots p_k^{\alpha_k},$$

where $p_1, \ldots, p_k$ are distinct primes and $\alpha_1, \ldots, \alpha_k$ are positive integers. This representation is called the **canonical factorization** (or **factorization**) of n. It is not difficult to see that the canonical factorization of the product of two integers is the product of the canonical factorizations of the two integers. This factorization allows us to establish the following fundamental property of primes.

Corollary 1.5. Let a and b be integers. If a prime p divides ab, then p divides either a or b.

Proof: Because p divides ab, p must appear in the canonical factorization of ab. The canonical factorizations of a, b, and ab are unique, and the canonical factorization of ab is the product of the canonical factorizations of a and b. Thus p must appear in at least one of the canonical factorizations of a and b, implying the desired result. □

Another immediate application of the prime factorization theorem is an alternative way of proving that there are infinitely many primes.

As in the proof of Theorem 1.3, assume that there are only finitely many primes: $p_1 < p_2 < \cdots < p_m$. Let

$$N = \prod_{i=1}^{m} \left(1 + \frac{1}{p_i} + \frac{1}{p_i^2} + \cdots\right) = \prod_{i=1}^{m} \frac{1}{1 - \frac{1}{p_i}}.$$

On the other hand, by expanding and by using the canonical factorization of positive integers, we obtain

$$N = 1 + \frac{1}{2} + \frac{1}{3} + \cdots,$$

yielding

$$\prod_{i=1}^{m} \frac{p_i}{p_i - 1} = \infty,$$

a contradiction. We have used the well-known facts:

(a) the harmonic series

$$1 + \frac{1}{2} + \frac{1}{3} + \cdots$$

diverges;

(b) the expansion formula

$$\frac{1}{1-x} = 1 + x + x^2 + \cdots$$

holds for real numbers x with $|x| < 1$. This expansion formula can also be interpreted as the summation formula for the infinite geometric progression $1, x, x^2, \ldots$.

From the formula

$$\prod_{i=1}^{\infty} \frac{p_i}{p_i - 1} = \infty,$$

using the inequality $1 + t \le e^t$, $t \in \mathbb{R}$, we can easily derive

$$\sum_{i=1}^{\infty} \frac{1}{p_i} = \infty.$$

For a prime p we say that p^k **fully divides** n and write $p^k \| n$ if k is the greatest positive integer such that $p^k | n$.

Example 1.9. [ARML 2003] Find the largest divisor of 1001001001 that does not exceed 10000.

Solution: We have

$$1001001001 = 1001 \cdot 10^6 + 1001 = 1001 \cdot (10^6 + 1) = 7 \cdot 11 \cdot 13 \cdot (10^6 + 1).$$

Note that $x^6 + 1 = (x^2)^3 + 1 = (x^2 + 1)(x^4 - x^2 + 1)$. We conclude that $10^6 + 1 = 101 \cdot 9901$, and so $1001001001 = 7 \cdot 11 \cdot 13 \cdot 101 \cdot 9901$. It is not difficult to check that no combination of 7, 11, 13, and 101 can generate a product greater than 9901 but less than 10000, so the answer is 9901. □

Example 1.10. Find n such that $2^n \| 3^{1024} - 1$.

Solution: The answer is 12.

Note that $2^{10} = 1024$ and $x^2 - y^2 = (x + y)(x - y)$. We have

$$\begin{aligned} 3^{2^{10}} - 1 &= (3^{2^9} + 1)(3^{2^9} - 1) = (3^{2^9} + 1)(3^{2^8} + 1)(3^{2^8} - 1) \\ &= \cdots = (3^{2^9} + 1)(3^{2^8} + 1)(3^{2^7} + 1) \cdots (3^{2^1} + 1)(3^{2^0} + 1)(3 - 1). \end{aligned}$$

By Example 1.5, $2 \| 3^{2^k} + 1$, for positive integers k. Thus the answer is $9 + 2 + 1 = 12$. □

Theorem 1.4 indicates that all integers are generated (productively) by primes. Because of the importance of primes, many people have tried to find (explicit) formulas to generate primes. So far, all the efforts are incomplete. On the other hand, there are many negative results. The following is a typical one, due to Goldbach:

Theorem 1.3b. *For any given integer m, there is no polynomial $p(x)$ with integer coefficients such that $p(n)$ is prime for all integers n with $n \geq m$.*

Proof: For the sake of contradiction, assume that there is such a polynomial

$$p(x) = a_k x_k + a_{k-1}x^{k-1} + \cdots + a_1 x + a_0$$

with $a_k, a_{k-1}, \ldots, a_0$ being integers and $a_k \neq 0$.

If $p(m)$ is composite, then our assumption was wrong. If not, assume that $p(m) = p$ is a prime. Then

$$p(m) = a_k m^k + a_{k-1}m^{k-1} + \cdots + a_1 m + a_0$$

and for positive integers i,

$$p(m+pi) = a_k(m+pi)^k + a_{k-1}(m+pi)^{k-1} + \cdots + a_1(m+pi) + a_0.$$

Note that

$$(m+pi)^j = m^j + \binom{j}{i} m^{j-1}(pi) + \binom{j}{2} m^{j-2}(pi)^2 + \cdots + \binom{j}{j-1} m(pi)^{j-1} + (pi)^j.$$

Hence $(m+pj)^j - m^j$ is a multiple of p. It follows that $p(m+pi) - p(m)$ is a multiple of p. Because $p(m) = p$, $p(m+pi)$ is a multiple of p. By our assumption, $p(m+pi)$ is also prime. Thus, the possible values of $p(m+pi)$ are 0, p, and $-p$ for all positive integers i. On the other hand, the equations $p(x) = 0$, $p(x) = p$, and $p(x) = -p$ can have at most $3k$ roots. Therefore, there exist (infinitely many) i such that $m + pi$ is not a solution of any of the equations $p(x) = 0$, $p(x) = p$, and $p(x) = -p$. We obtain a contradiction. Hence our assumption was wrong. Therefore, such polynomials do not exist. □

Even though there are no definitive ways to find primes, the density of primes (that is, the average appearance of primes among integers) has been known for about 100 years. This was a remarkable result in the mathematical field of *analytic number theory* showing that

$$\lim_{n\to\infty} \frac{\pi(n)}{n/\log n} = 1,$$

where $\pi(n)$ denotes the number of primes $\leq n$. The relation above is known as the prime number theorem. It was proved by Hadamard and de la Vallée Poussin in 1896. An elementary but difficult proof was given by Erdös and Selberg.

G.C.D.

For a positive integer k we denote by D_k the set of all its positive divisors. It is clear that D_k is a finite set. For positive integers m and n the maximal element in the set $D_m \cap D_n$ is called the **greatest common divisor** (or **G.C.D.**) of m and n and is denoted by $\gcd(m, n)$. In the case $D_m \cap D_n = \{1\}$, we have $\gcd(m, n) = 1$ and we say that m and n are **relatively prime** (or **coprime**). The following are some basic properties of G.C.D.

Proposition 1.6.

(a) if p is a prime, then $\gcd(p, m) = p$ or $\gcd(p, m) = 1$.

(b) If $d = \gcd(m, n)$, $m = dm'$, $n = dn'$, then $\gcd(m', n') = 1$.

(c) If $d = \gcd(m, n)$, $m = d'm''$, $n = d'n''$, $\gcd(m'', n'') = 1$, then $d' = d$.

(d) If d' is a common divisor of m and n, then d' divides $\gcd(m, n)$.

(e) If $p^x \| m$ and $p^y \| n$, then $p^{\min x, y} \| \gcd(m, n)$. Furthermore, if $m = p_1^{\alpha_1} \cdots p_k^{\alpha_k}$ and $n = p_1^{\beta_1} \cdots p_k^{\beta_k}$, $\alpha_i, \beta_i \geq 0$, $i = 1, \ldots, k$, then

$$\gcd(m, n) = p_1^{\min(\alpha_1, \beta_1)} \cdots p_k^{\min(\alpha_k, \beta_k)}.$$

(f) If $m = nq + r$, then $\gcd(m, n) = \gcd(n, r)$.

Proof: The proofs of these properties are rather straightforward from the definition. We present only the proof property (f). Set $d = \gcd(m, n)$ and $d' = \gcd(n, r)$. Because $d \mid m$ and $d \mid n$ it follows that $d \mid r$. Hence $d \mid d'$. Conversely, from $d' \mid n$ and $d' \mid r$ it follows that $d' \mid m$, so $d' \mid d$. Thus $d = d'$. □

The definition of G.C.D. can easily be extended to more than two numbers. For given integers $a_1, a_2, \ldots, a_n$, $\gcd(a_1, a_2, \ldots, a_n)$ is the common greatest divisor of all the numbers $a_1, a_2, \ldots, a_n$. We can define the greatest common divisor of $a_1, a_2, \ldots, a_n$ by considering

$$d_1 = \gcd(a_1, a_2),\ d_2 = \gcd(d_1, a_3), \ldots, d_{n-1} = \gcd(d_{n-2}, a_n).$$

We leave to the reader to convince himself that $d_{n-1} = \gcd(a_1, \ldots, a_n)$. We also leave the simple proofs of the following properties to the reader.

Proposition 1.6. (Continuation)

(g) $\gcd(\gcd(m, n), p) = \gcd(m, \gcd(n, p))$; proving that $\gcd(m, n, p)$ is well-defined;

(h) If $d \mid a_i$, $i = 1, \ldots, s$, then $d \mid \gcd(a_1, \ldots, a_s)$;

(i) If $a_i = p_1^{\alpha_{1i}} \cdots p_k^{\alpha_{ki}}$, $i = 1, \ldots, s$, then

$$\gcd(a_1, \ldots, a_s) = p_1^{\min(\alpha_{11}, \ldots, \alpha_{1k})} \cdots p_k^{\min(\alpha_{k1}, \ldots, \alpha_{kk})}.$$

We say that $a_1, a_2, \ldots, a_n$ are relatively prime if their greatest common divisor is equal to 1. Note that $\gcd(a_1, a_2, \ldots, a_n) = 1$ does not imply that $\gcd(a_i, a_j) = 1$ for $1 \le i < j \le n$. (For example, we can set $a_1 = 2, a_2 = 3$, and $a_3 = 6$.) If $a_1, a_2, \ldots, a_n$ are such that $\gcd(a_i, a_j) = 1$ for $1 \le i < j \le n$, we say that these numbers are pairwise relatively prime (or coprime).

Euclidean Algorithm

Canonical factorizations help us to determine the greatest common divisors of integers. But it is not easy to factor numbers, especially large numbers. (This is why we need to study divisibility of numbers.) A useful algorithm for finding the greatest common divisor of two positive integers m and n is the **Euclidean algorithm**. It consists of repeated application of the division algorithm:

$$\begin{aligned} m &= nq_1 + r_1, \quad 1 \le r_1 < n, \\ n &= r_1 q_2 + r_2, \quad 1 \le r_2 < r_1, \\ &\vdots \\ r_{k-2} &= r_{k-1} q_k + r_k, \quad 1 \le r_k < r_{k-1}, \\ r_{k-1} &= r_k q_{k+1} + r_{k+1}, \quad r_{k+1} = 0. \end{aligned}$$

This chain of equalities is finite because $n > r_1 > r_2 > \cdots > r_k$.

The last nonzero remainder, r_k, is the greatest common divisor of m and n. Indeed, by applying successively property (f) above we obtain

$$\gcd(m, n) = \gcd(n, r_1) = \gcd(r_1, r_2) = \cdots = \gcd(r_{k-1}, r_k) = r_k.$$

Example 1.11. [HMMT 2002] If a positive integer multiple of 864 is chosen randomly, with each multiple having the same probability of being chosen, what is the probability that it is divisible by 1944?

First Solution: The probability that a multiple of $864 = 2^5 \cdot 3^3$ is divisible by $1944 = 2^3 \cdot 3^5$ is the same as the probability that a multiple of $2^2 = 4$ is divisible by $3^2 = 9$. Since 4 and 9 are relatively prime, the probability is $\frac{1}{9}$. □

Second Solution: By the Euclidean algorithm, we have $\gcd(1944, 864) = \gcd(1080, 864) = \gcd(864, 216) = 216$. Hence $1944 = 9 \cdot 216$ and $864 = 4 \cdot 216$. We can finish as in the first solution. □

Example 1.12. [HMMT 2002] Compute

$$\gcd(2002+2, 2002^2+2, 2002^3+2, \ldots).$$

Solution: Let g denote the desired greatest common divisor. Note that $2002^2 + 2 = 2002(2000+2)+2 = 2000(2002+2)+6$. By the Euclidean algorithm, we have

$$\gcd(2002+2, 2002^2+2) = \gcd(2004, 6) = 6.$$

Hence $g \mid \gcd(2002+2, 2002^2+2) = 6$. On the other hand, every number in the sequence $2002+2, 2002^2+2, \ldots$ is divisible by 2. Furthermore, since $2002 = 2001 + 1 = 667 \cdot 3 + 1$, for all positive integers k, $2002^k = 3a_k + 1$ for some integer a_k. Thus $2002^k + 2$ is divisible by 3. Because 2 and 3 are relatively prime, every number in the sequence is divisible by 6. Therefore, $g = 6$. □

Bézout's Identity

Let's start with two classic brain teasers.

Example 1.13. In a special football game, a team scores 7 points for a touchdown and 3 points for a field goal. Determine the largest mathematically unreachable number of points scored by a team in an (infinitely long) game.

Solution: The answer is 11. It's not difficult to check that we cannot obtain 11 points. Note that $12 = 3+3+3+3$, $13 = 7+3+3$, and $14 = 7+7$. For all integers n greater than 11, the possible remainders when n is divided by 3 are 0, 1, and 2. If n has remainder 0, we can clearly obtain n points by scoring enough field goals; if n has remainder 1, then $n - 13$ has remainder 0, and we can obtain n points by scoring one touchdown and enough field goals; if n has remainder 2, then $n - 14$ has remainder 0, and we can obtain n points by scoring two touchdowns and enough field goals. In short, all integers n greater than 11 can be written in the form $n = 7a + 3b$ for some nonnegative integers a and b. □

Example 1.14 There is an ample supply of milk in a milk tank. Mr. Fat is given a 5-liter (unmarked) container and a 9-liter (unmarked) container. How can he measure out 2 liters of milk?

Solution: Let T, L_5, and L_9 denote the milk tank, the 5-liter container, and the 9-liter container, respectively. We can use the following table to achieve the desired result.

T	L_5	L_9
x	0	0
$x-5$	5	0
$x-5$	0	5
$x-10$	5	5
$x-10$	1	9
$x-1$	1	0
$x-1$	0	1
$x-6$	5	1
$x-6$	0	6
$x-11$	5	6
$x-11$	**2**	9

□

The key is to make the connection between $2 = 4 \times 5 - 2 \times 9$. We leave it to the reader to use the equation $2 = 3 \times 9 - 5 \times 5$ to set up another process. For given integers $a_1, a_2, \ldots, a_n$, we call $\alpha_1 a_1 + \alpha_2 a_2 + \cdots + \alpha_n a_n$, where $\alpha_1, \alpha_2, \ldots, \alpha_n$ are arbitrary integers, **linear combinations** of $a_1, a_2, \ldots, a_n$. Examples 1.13 and 1.14 are seemingly unrelated problems. But they both involve linear combinations of two given integers. What if we replace (7, 3) by (6, 3) in Example 1.13, and (5, 9) by (6, 9) in Example 1.14? We have the following general result.

Theorem 1.7. [Bézout] For positive integers m and n, there exist integers x and y such that $mx + ny = \gcd(m, n)$.

Proof: From the Euclidean algorithm it follows that

$$r_1 = m - nq_1, \quad r_2 = -mq_2 + n(1 + q_1 q_2), \ldots.$$

In general, $r_i = m\alpha_i + n\beta_i$, for $i = 1, \ldots, k$. Because $r_{i+1} = r_{i-1} - r_i q_{i+1}$, it follows that

$$\alpha_{i+1} = \alpha_{i-1} - q_{i+1}\alpha_i,$$
$$\beta_{i+1} = \beta_{i-1} - q_{i+1}\beta_i,$$

for $i = 2, \ldots, k-1$. Finally, we obtain $\gcd(m, n) = r_k = \alpha_k m + \beta_k n$. □

Note that $\gcd(a, b)$ divides $ax + by$. In view of Bézout's identity, for given integers a, b, and c, the equation $ax + by = c$ is solvable for integers (x, y) if and only if $\gcd(a, b)$ divides c. In algebra, we solve systems of equations. In number theory, we usually try to find special solutions for systems of equations, namely, integer solutions, rational solutions, and so on. Hence most of the these systems have more variables than the number of equations in the system. These are called **Diophantine equations**, attributed to the ancient Greek mathematician Diophantus, which will be studied extensively in the sequel to this book: *105 Diophantine Equations and Integer Function Problems*. For fixed integers a, b, and c, $ax + by = c$ is a two-variable linear Diophantine equation.

Corollary 1.8. If $a \mid bc$ and $\gcd(a, b) = 1$, then $a \mid c$.

Proof: If $c = 0$, the assertion is clearly true. Assume that $c \neq 0$. Since $\gcd(a, b) = 1$, by Bézout's identity, $ax + by = 1$ for some integers x and y. Hence $acx + bcy = c$. Because a divides acx and bcy, a divides c, as desired. □

Corollary 1.9. Let a and b be two coprime numbers. If c is an integer such that $a \mid c$ and $b \mid c$, then $ab \mid c$.

Proof: Because $a \mid c$, we have $c = ax$ for some integer x. Hence b divides ax. Because $\gcd(a, b) = 1$, $b \mid x$, and by Corollary 1.8, it follows that $x = by$ for some integer y, and so $c = aby$, or $ab \mid c$. □

Corollary 1.10. Let p be a prime, and let k be an integer with $1 \leq k < p$. Then $p \mid \binom{p}{k}$.

Proof: Note that from relation

$$k\binom{p}{k} = p\binom{p-1}{k-1}$$

it follows that p divides $k\binom{p}{k}$. Because $gcd(p, k) = 1$, the relation p divides $\binom{p}{k}$ is obtained via Corollary 1.8. □

Example 1.15. [Russia 2001] Let a and b be distinct positive integers such that $ab(a + b)$ is divisible by $a^2 + ab + b^2$. Prove that $|a - b| > \sqrt[3]{ab}$.

Proof: Set $g = \gcd(a, b)$ and write $a = xg$ and $b = yg$ with $\gcd(x, y) = 1$. Then

$$\frac{ab(a+b)}{a^2 + ab + b^2} = \frac{xy(x+y)g}{x^2 + xy + y^2}$$

is an integer. Note that $\gcd(x^2 + xy + y^2, x) = \gcd(y^2, x) = 1$. Similarly, $\gcd(x^2 + xy + y^2, y) = 1$. Because $\gcd(x + y, y) = 1$, we have

$$\gcd(x^2 + xy + y^2, x + y) = \gcd(y^2, x + y) = 1.$$

By Corollary 1.9,

$$x^2 + xy + y^2 \mid g,$$

implying that $g \geq x^2 + xy + y^2$. Therefore,

$$\begin{aligned} |a - b|^3 &= |g(x - y)|^3 = g^2|x - y|^3 \cdot g \\ &\geq g^2 \cdot 1 \cdot (x^2 + xy + y^2) \\ &> g^2 xy = ab. \end{aligned}$$

It follows that $|a - b| > \sqrt[3]{ab}$. □

Note that the key step $x^2 + xy + y^2$ divides g can also be obtained by clever algebraic manipulations such as $a^3 = (a^2 + ab + b^2)a - ab(a + b)$.

L.C.M.

For a positive integer k we denote by M_k the set of all multiples of k. As opposed to the set D_k defined earlier in this section, M_k is an infinite set.

For positive integers s and t the minimal element of the set $M_s \cap M_t$ is called the *least common multiple* of s and t and is denoted by $\operatorname{lcm}(s,t)$ or $[s,t]$.

Proposition 1.11.

(a) If $\operatorname{lcm}(s,t) = m$, $m = ss' = tt'$, then $\gcd(s',t') = 1$.

(b) If m' is a common multiple of s and t and $m' = ss' = tt'$, $\gcd(s',t') = 1$, then $m' = m$.

(c) If m' is a common multiple of s and t, then $m \mid m'$.

(d) If $m \mid s$ and $n \mid s$, then $\operatorname{lcm}(m,n) \mid s$.

(e) If n is an integer, $n \operatorname{lcm}(s,t) = \operatorname{lcm}(ns,nt)$.

(f) If $s = p_1^{\alpha_1} \cdots p_k^{\alpha_k}$ and $t = p_1^{\beta_1} \cdots p_k^{\beta_k}$, $\alpha_i, b_i \geq 0$, $i = 1, \ldots, k$, then

$$\operatorname{lcm}(s,t) = p_1^{\max(\alpha_1,\beta_1)} \cdots p_k^{\max(\alpha_k,\beta_k)}.$$

The properties in Proposition 1.11 are easily obtained from the definition of L.C.M., and we leave their proofs to the reader.

The following property establishes an important connection between G.C.D. and L.C.M.

Proposition 1.12. For any positive integers m and n the following relation holds:

$$mn = \gcd(m,n) \cdot \operatorname{lcm}(m,n).$$

Proof: Let $m = p_1^{\alpha_1} \cdots p_k^{\alpha_k}$, $n = p_1^{\beta_1} \cdots p_k^{\beta_k}$, $\alpha_i, \beta_i \geq 0$, $i = 1, \ldots, k$. From Properties 1.6 (e) and 1.11 (f) we have

$$\begin{aligned}\gcd(m,n)\operatorname{lcm}(m,n) &= p_1^{\min(\alpha_1,\beta_1)+\max(\alpha_1,\beta_1)} \cdots p_k^{\min(\alpha_k,\beta_k)+\max(\alpha_k,\beta_k)} \\ &= p_1^{\alpha_1+\beta_1} \cdots p_k^{\alpha_k+\beta_k} = mn. \qquad \square\end{aligned}$$

Let $a_1, a_2, \ldots, a_n$ be positive integers. The **least common multiple** of $a_1, a_2, \ldots, a_n$, denoted by $\operatorname{lcm}(a_1, a_2, \ldots, a_n)$, is the smallest positive integer that is a multiple of all of $a_1, a_2, \ldots, a_n$. Note that Proposition 1.12 cannot be easily generalized. For example, it is not true that

$$\gcd(a,b,c)\operatorname{lcm}(a,b,c) = abc.$$

We leave it to the reader to find interesting counterexamples.

The Number of Divisors

We start with three examples.

Example 1.16. [AIME 1988] Compute the probability that a randomly chosen positive divisor of 10^{99} is an integer multiple of 10^{88}.

Solution: What are the divisors of 10^{99}? Is 3 a divisor? Is 220 a divisor? We consider the prime factorization of 10^{99}, which is $2^{99} \cdot 5^{99}$. The divisors of 10^{99} are of the form $2^a \cdot 5^b$, where a and b are integers with $0 \le a, b \le 99$. Because there are 100 choices for each of a and b, 10^{99} has $100 \cdot 100$ positive integer divisors. Of these, the multiples of $10^{88} = 2^{88} \cdot 5^{88}$ must satisfy the inequalities $88 \le a, b \le 99$. Thus there are 12 choices for each of a and b; that is, $12 \cdot 12$ of the $100 \cdot 100$ divisors of 10^{99} are multiples of 10^{88}. Consequently, the desired probability is $\frac{12\cdot 12}{100\cdot 100} = \frac{9}{625}$. □

Example 1.17. Determine the number of ordered pairs of positive integers (a, b) such that the least common multiple of a and b is $2^3 5^7 11^{13}$.

Solution: Both a and b are divisors of $2^3 5^7 11^{13}$, and so $a = 2^x 5^y 11^z$ and $b = 2^s 5^t 11^u$ for some nonnegative integers x, y, z, s, t, u. Because $2^3 5^7 11^{13}$ is the least common multiple, $\max\{x, s\} = 3$, $\max\{y, t\} = 7$, and $\max\{z, u\} = 13$. Hence (x, s) can be equal to $(0, 3)$, $(1, 3)$, $(2, 3)$, $(3, 3)$, $(3, 2)$, $(3, 1)$, or $(3, 0)$, so there are 7 choices for (x, s). Similarly, there are 15 and 27 choices for (y, t) and (z, u), respectively. By the multiplication principle, there are $7 \times 15 \times 27 = 2835$ ordered pairs of positive integers (a, b) having $2^3 5^7 11^{13}$ as their least common multiple. □

Example 1.18. Determine the product of distinct positive integer divisors of $n = 420^4$.

Solution: Because $n = (2^2 \cdot 3 \cdot 5 \cdot 7)^4$, d is a divisor of n if and only if d can be written in the form $2^a \cdot 3^b \cdot 5^c \cdot 7^d$, where $0 \le a \le 8$, $0 \le b \le 4$, $0 \le c \le 4$, and $0 \le d \le 4$. Hence there are 9, 5, 5, and 5 possible values for a, b, c, and d, respectively. It follows that n has $9\cdot 5\cdot 5\cdot 5 = 1125$ positive divisors. If $d \ne 420^2$, then $\frac{420^4}{d}$ is also a divisor, and the product of these two divisors is 420^4. We can thus partition 1124 divisors of n (excluding 420^2) into 562 pairs of divisors of the form $\left(d, \frac{n}{d}\right)$, and the product of the two divisors in each pair is 420^4. Hence the answer is

$$420^{4\cdot 562} \cdot 420^2 = 420^{2250}. \qquad \square$$

Putting the last three examples together gives two interesting results in number theory. For a positive integer n denote by $\tau(n)$ the number of its divisors. It is clear that

$$\tau(n) = \sum_{d|n} 1.$$

Writing τ in this summation form allows us later to discuss it as an example of a **multiplicative arithmetic function**.

Proposition 1.13. If $n = p_1^{a_1} p_2^{a_2} \cdots p_k^{a_k}$ is a prime decomposition of n, then n has

$$\tau(n) = (a_1 + 1)(a_2 + 1) \cdots (a_k + 1) \text{ divisors.}$$

Corollary 1.14. If $n = p_1^{a_1} p_2^{a_2} \cdots p_k^{a_k}$ is a prime decomposition of n, then there are

$$(2a_1 + 1)(2a_2 + 1) \cdots (2a_k + 1)$$

distinct pairs of ordered positive integers (a, b) with $\text{lcm}(a, b) = n$.

Corollary 1.15. For any positive integer n,

$$\prod_{d|n} d = n^{\frac{\tau(n)}{2}}.$$

The proofs of these three propositions are identical to those of Examples 1.16, 1.17, and 1.18. It is interesting to note that these three results can be generalized to the case that the powers of the primes in the prime decomposition are nonnegative (because if $a_i = 0$ for some $1 \le i \le k$, then $a_i + 1 = 2a_i + 1 = 1$, which does not affect the products).

Corollary 1.16. For any positive integer n, $\tau(n) \le 2\sqrt{n}$.

Proof: Let $d_1 < d_2 < \cdots < d_k$ be the divisors of n not exceeding $\sqrt{n}$. The remaining divisors are

$$\frac{n}{d_1}, \frac{n}{d_2}, \ldots, \frac{n}{d_k}.$$

It follows that $\tau(n) \le 2k \le 2\sqrt{n}$. □

The Sum of Divisors

For a positive integer n denote by $\sigma(n)$ the sum of its positive divisors, including 1 and n itself. It is clear that

$$\sigma(n) = \sum_{d|n} d.$$

This representation will help us to show that σ is multiplicative.

Proposition 1.17. If $n = p_1^{\alpha_1} \cdots p_k^{\alpha_k}$ is the prime factorization of n, then

$$\sigma(n) = \frac{p_1^{\alpha_1+1} - 1}{p_1 - 1} \cdots \frac{p_k^{\alpha_k+1} - 1}{p_k - 1}.$$

Proof: The divisors of n can be written in the form

$$p_1^{a_1} \cdots p_k^{a_k},$$

where $a_1, \ldots, a_k$ are integers with $0 \le a_1 \le \alpha_1, \ldots, 0 \le a_k \le \alpha_k$. Each divisor of n appears exactly once as a summand in the expansion of the product

$$(1 + p_1 + \cdots + p_1^{\alpha_1}) \cdots (1 + p_k + \cdots + p_k^{\alpha_k}),$$

from which the desired result follows, by also noting the formula for the sum of a finite geometric progression:

$$\frac{r^{k+1} - 1}{r - 1} = 1 + r + r^2 + \cdots + r^k. \qquad \square$$

Example 1.19. Find the sum of even positive divisors of 10000.

Solution: The even divisors of 10000 can be written in the form of $2^a 5^b$, where a and b are integers with $1 \le a \le 5$ and $0 \le b \le 5$. Each even divisor of 10000 appears exactly once as a summand in the expansion of the product

$$\begin{aligned}(2 + 2^2 + 2^3 + 2^4 + 2^5)(1 + 5 + 5^2 + 5^3 + 5^4 + 5^5) &= 62 \cdot \frac{5^6 - 1}{5 - 1} \\ &= 242172. \qquad \square\end{aligned}$$

Modular Arithmetic

Let a, b, and m be integers, with $m \ne 0$. We say that a and b are **congruent modulo** m if m divides $a - b$. We denote this by $a \equiv b \pmod{m}$. The relation "$\equiv$" on the set $\mathbb{Z}$ of integers is called the **congruence relation**. If m does not divide $a - b$, then we say that integers a and b are not congruent modulo m and we write $a \not\equiv b \pmod{m}$.

Proposition 1.18.

(a) $a \equiv a \pmod{m}$ (reflexivity).

(b) If $a \equiv b \pmod{m}$ and $b \equiv c \pmod{m}$, then $a \equiv c \pmod{m}$ (transitivity).

(c) If $a \equiv b \pmod{m}$, then $b \equiv a \pmod{m}$.

(d) If $a \equiv b \pmod{m}$ and $c \equiv d \pmod{m}$, then $a + c \equiv b + d \pmod{m}$ and $a - c \equiv b - d \pmod{m}$.

(e) If $a \equiv b \pmod{m}$, then for any integer k, $ka \equiv kb \pmod{m}$.

(f) If $a \equiv b \pmod{m}$ and $c \equiv d \pmod{m}$, then $ac \equiv bd \pmod{m}$. In general, if $a_i \equiv b_i \pmod{m}$, $i = 1, \ldots, k$, then $a_1 \cdots a_k \equiv b_1 \cdots b_k \pmod{m}$. In particular, if $a \equiv b \pmod{m}$, then for any positive integer k, $a^k \equiv b^k \pmod{m}$.

(g) We have $a \equiv b \pmod{m_i}$, $i = 1, \ldots, k$, if and only if

$$a \equiv b \pmod{\operatorname{lcm}(m_1, \ldots, m_k)}.$$

In particular, if $m_1, \ldots, m_k$ are pairwise relatively prime, then $a \equiv b \pmod{m_i}$, $i = 1, \ldots, k$, if and only if $a \equiv b \pmod{m_1 \cdots m_k}$.

Proof: The proofs are straightforward. We present the proof of (g) and leave the rest to the reader.

From $a \equiv b \pmod{m_i}$, $i = 1, \ldots, k$, it follows that $m_i \mid (a - b)$, $i = 1, \ldots, k$. Hence $a - b$ is a common multiple of $m_1, \ldots, m_k$, and so $\operatorname{lcm}(m_1, \ldots, m_k) \mid (a - b)$. That is, $a \equiv b \pmod{\operatorname{lcm}(m_1, \ldots, m_k)}$.

Conversely, from $a \equiv b \pmod{\operatorname{lcm}(m_1, \ldots, m_k)}$ and the fact that each m_i divides $\operatorname{lcm}(m_1, \ldots, m_k)$ we obtain $a \equiv b \pmod{m_i}$, $i = 1, \ldots, k$. □

Proposition 1.19. Let a, b, n be integers, $n \neq 0$, such that $a = nq_1 + r_1$, $b = nq_2 + r_2$, $0 \leq r_1, r_2 < |n|$. Then $a \equiv b \pmod{n}$ if and only if $r_1 = r_2$.

Proof: Because $a - b = n(q_1 - q_2) + (r_1 - r_2)$, it follows that $n \mid (a - b)$ if and only if $n \mid (r_1 - r_2)$. Taking into account that $|r_1 - r_2| < |n|$, we have $n \mid (r_1 - r_2)$ if and only if $r_1 = r_2$. □

Example 1.20. Prove that there are infinitely many primes of the form $4k - 1$; that is, congruent to 3 modulo 4.

Proof: We first note that there is at least one prime p with $p \equiv 3 \pmod{4}$ (simply set $p = 3$). Suppose there were only finitely many primes congruent to 3 modulo 4. Let $p_1, p_2, \ldots, p_k$ be those primes, and let $P = p_1 p_2 \cdots p_k$ denote their product. We have $4P - 1 \equiv 3 \pmod{4}$. If all the prime divisors of $4P - 1$ were congruent to 1 modulo 4, then $4P - 1$ would be congruent to 1 modulo 4 (by Proposition 1.18 (g)). Thus, some prime divisor p of $4P - 1$ would be congruent to 3 modulo 4. On the other hand, $\gcd(4P - 1, p_i) = 1$ for all i with $1 \leq i \leq k$, and so we find another prime that is congruent to 3 modulo 4, a contradiction to our assumption. Hence there are infinitely many primes of the form $4k - 1$. □

In exactly the same way, we can show that there are infinitely many primes of the form $6k - 1$. We can view congruency as (part of) an arithmetic progression. For example, we can rewrite the last two results as follows: There are infinitely many primes in the arithmetic progression $\{-1 + ka\}_{k=1}^{\infty}$ with $a = 4$ or $a = 6$. These are the special cases of a famous result of Dirichlet:

> There are infinitely many primes in any arithmetic progression of integers for which the common difference is relatively prime to the terms. In other words, if a and m be relatively prime positive integers, then there are infinitely many primes p such that $p \equiv a \pmod{m}$.

Dirichlet was also able to compute the density (in simpler terms, a certain kind of frequency of such primes) of these prime numbers in the set of all primes. This was another milestone in analytic number theory. The proof of this work is beyond the scope of this book. We present a more detailed form of this result in the glossary section of this book. Some problems in this book become easy if we apply this theorem directly. But all of these problems can also be solved in different ways, and we strongly encourage the reader to look for these different approaches, which will enhance the reader's problem-solving abilities.

In Example 1.20, it is very natural to work modulo 4. Many times, such a choice is not obvious. Taking the proper modulus holds the key to many problems.

Example 1.21. [Russia 2001] Find all primes p and q such that $p + q = (p - q)^3$.

Solution: The only such primes are $p = 5$ and $q = 3$.

Because $(p - q)^3 = p + q \neq 0$, p and q are distinct and hence relatively prime.

Because $p - q \equiv 2p \pmod{p + q}$, taking the given equation modulo $p + q$ gives $0 \equiv 8p^3 \pmod{p + q}$. Because p and q are relatively prime, so are p and $p + q$. Thus, $0 \equiv 8 \pmod{p + q}$; that is, $p + q$ divides 8.

Likewise, taking the given equation modulo $p - q$ gives $2p \equiv 0 \pmod{p - q}$. Because p and q are relatively prime, so are p and $p - q$. We conclude that $2 \equiv 0 \pmod{p - q}$, or $p - q$ divides 2.

It easily follows that (p, q) is equal to $(3, 5)$ or $(5, 3)$; only the latter satisfies the given equation. □

There is another approach to the last problem: setting $p - q = a$ leads to $p + q = a^3$. Hence $p = \frac{a^3 + a}{2}$ and $q = \frac{a^3 - a}{2}$. This kind of substitution is a very common technique in solving Diophantine equations.

Example 1.22. [Baltic 2001] Let a be an odd integer. Prove that $a^{2^n} + 2^{2^n}$ and $a^{2^m} + 2^{2^m}$ are relatively prime for all positive integers n and m with $n \neq m$.

Proof: Without loss of generality, assume that $m > n$. For any prime p dividing $a^{2^n} + 2^{2^n}$, we have

$$a^{2^n} \equiv -2^{2^n} \pmod{p}.$$

We square both sides of the equation $m - n$ times to obtain

$$a^{2^m} \equiv 2^{2^m} \pmod{p}.$$

Because a is odd, we have $p \neq 2$. Thus, $2^{2^m} + 2^{2^m} = 2^{2^m+1} \not\equiv 0 \pmod{p}$, so that

$$a^{2^m} \equiv 2^{2^m} \not\equiv -2^{2^m} \pmod{p}.$$

Therefore, $p \nmid (a^{2^m} + 2^{2^m})$, proving the desired result. □

Setting $a = 1$ in the last example leads to a property of the **Fermat numbers**, which will soon be discussed.

Example 1.23. Determine whether there exist infinitely many even positive integers k such that for every prime p the number $p^2 + k$ is composite.

Solution: The answer is positive.

First note that for $p = 2$, $p^2 + k$ is always composite for all even positive integers k.

Next we note that if $p > 3$, then $p^2 \equiv 1 \pmod{3}$. Hence if k is an even positive integer with $k \equiv 2 \pmod{3}$, then $p^2 + k$ is composite for all all primes $p > 3$ ($p^2 + k$ is greater than 3 and is divisible by 3).

Finally, we note that $3^2 + k \equiv 0 \pmod{5}$ if $k \equiv 1 \pmod{5}$.

Putting the above arguments together, we conclude that all positive integers k with

$$\begin{cases} k \equiv 0 \pmod{2}, \\ k \equiv 2 \pmod{3}, \\ k \equiv 1 \pmod{5}, \end{cases} \qquad (*)$$

satisfy the conditions of the problem. By Proposition 1.18 (g), we consider (mod lcm$(2, 3, 5)$) = (mod 30). It is not difficult to check that all positive integers k with $k = 26 \pmod{30}$ satisfy the system, and hence the conditions of the problem. □

The system $(*)$ is a **linear congruence system**, and each of the three equations in the system is a **linear congruence equation**. We will study the solutions of the linear congruence systems when we study the **Chinese Remainder Theorem** in the sequel to this book: *105 Diophantine Equations and Integer Function*

Problems. The major difference between solving an equation and solving a congruence equation is the limitation of division in the latter situation. For example, in algebra, $4x = 4y$ implies that $x = y$. In modular arithmetic, $4x \equiv 4y \pmod 6$ does not necessarily imply that $x \equiv y \pmod 6$. (Why?) On the other hand, $4x \equiv 4y \pmod{15}$ does imply that $x \equiv y \pmod{15}$. (Why?) Proposition 1.18 (g) plays a key role in this difference. In algebra, $xy = 0$ implies that either $x = 0$ or $y = 0$ or both. But in modular arithmetic, $xy \equiv 0 \pmod m$ does not imply $x \equiv 0 \pmod m$ or $y \equiv 0 \pmod m$. (For example, $3 \cdot 5 \equiv 0 \pmod{15}$, but $3 \not\equiv 0 \pmod{15}$ and $5 \not\equiv 0 \pmod{15}$. We will discuss this topic in a detailed fashion when we talk about linear congruence equations. For a little preview, we rewrite Corollary 1.5 in the language of modular arithmetic.

Corollary 1.20. Let p be a prime. If x and y are integers such that $xy \equiv 0 \pmod p$, then either $x \equiv 0 \pmod p$ or $y \equiv 0 \pmod p$ or both.

This is an example of interchanging the faces of a common idea in number theory: $p \mid xy$ (divisibility notation), $xy \equiv 0 \pmod p$ (modular and congruence notation), and $p = kxy$ (Diophantine equation forms). Simple applications Corollaries 1.8 and 1.9 also lead to the following properties.

Corollary 1.21. Let m be a positive integer, and let a, b, and c be integers with $c \neq 0$. If $ac \equiv bc \pmod m$, then

$$a \equiv b \left(\text{mod } \frac{m}{\gcd(c, m)}\right).$$

Corollary 1.22. Let m be a positive integer. Let a be an integer relatively prime to m. If a_1 and a_2 are integers such that $a_1 \not\equiv a_2 \pmod m$, then $a_1 a \not\equiv a_2 a \pmod m$.

The following property is useful in reducing the power in congruency relations.

Corollary 1.23. Let m be a positive integer, and let a and b be integers relatively prime to m. If x and y are integers such that

$$a^x \equiv b^x \pmod m \quad \text{and} \quad a^y \equiv b^y \pmod m,$$

then

$$a^{\gcd(x,y)} \equiv b^{\gcd(x,y)} \pmod m.$$

Proof: By Bézout's identity, there are nonnegative integers u and v such that $\gcd(x, y) = ux - vy$. By the given conditions, we have

$$a^{ux} \equiv b^{ux} \pmod m \quad \text{and} \quad b^{vy} \equiv a^{vy} \pmod m,$$

implying that $a^{ux}b^{vy} \equiv a^{vy}b^{ux} \pmod{m}$. Since $\gcd(a, m) = \gcd(b, m) = 1$, by Corollary 1.21, we have

$$a^{\gcd(x,y)} \equiv a^{ux-vy} \equiv b^{ux-vy} \equiv b^{\gcd(x,y)} \pmod{m}. \qquad \square$$

Residue Classes

By Proposition 1.18 (a), (b), and (c), we conclude that for any given positive integer m, we can classify integers into a unique class according to their remainder on division by m. Clearly, there are m such classes. A set S of integers is also called a **complete set of residue classes** modulo n if for each $0 \le i \le n-1$, there is an element $s \in S$ such that $i \equiv s \pmod{n}$. Clearly, $\{a, a+1, a+2, \ldots, a+m-1\}$ is a **complete set of residue classes** modulo m for any integer a. In particular, for $a = 0$, $\{0, 1, \ldots, m-1\}$ is the **minimal nonnegative complete set of residue classes**. Also, it is common to consider the complete set of residue classes $\{0, \pm 1, \pm 2, \ldots, \pm k\}$ for $m = 2k+1$ and $\{0, \pm 1, \pm 2, \ldots, \pm(k-1), k\}$ for $m = 2k$.

Example 1.24. Let n be an integer. Then

(1) $n^2 \equiv 0$ or $1 \pmod{3}$;

(2) $n^2 \equiv 0$ or $\pm 1 \pmod{5}$;

(3) $n^2 \equiv 0$ or 1 or $4 \pmod{8}$;

(4) $n^3 \equiv 0$ or $\pm 1 \pmod{9}$;

(5) $n^4 \equiv 0$ or $1 \pmod{16}$;

All the proofs can be done by checking complete sets of residue classes. We leave them to the reader. We also encourage the reader to review these relations after finishing studying Euler's theorem.

Example 1.25. [Romania 2003] Consider the prime numbers $n_1 < n_2 < \cdots < n_{31}$. Prove that if 30 divides $n_1^4 + n_2^4 + \cdots + n_{31}^4$, then among these numbers one can find three consecutive primes.

Solution: Let $s = n_1^4 + n_2^4 + \cdots + n_{31}^4$.

Firstly, we claim that $n_1 = 2$. Otherwise, all numbers n_i, $1 \le i \le 31$, are odd, and consequently s is odd, a contradiction.

Secondly, we claim that $n_2 = 3$. Otherwise, we have $n_i^4 \equiv 1 \pmod{3}$ for all $1 \le i \le 31$. It follows that $s \equiv 31 \equiv 1 \pmod{3}$, a contradiction.

Finally, we prove that $n_3 = 5$. Indeed, if not, then $n_i^2 \equiv \pm 1 \pmod{5}$ and $n_i^4 \equiv 1 \pmod{5}$ for all $1 \le i \le 31$. Thus, $s \equiv 31 \equiv 1 \pmod{5}$, a contradiction.

We conclude that three consecutive primes, namely, 2, 3, and 5, appear in the given prime numbers. $\square$

Example 1.26. Let m be an even positive integer. Assume that

$$\{a_1, a_2, \ldots, a_m\} \quad \text{and} \quad \{b_1, b_2, \ldots, b_m\}$$

are two complete sets of residue classes modulo m. Prove that

$$\{a_1 + b_1, a_2 + b_2, \ldots, a_m + b_m\}$$

is not a complete set of residue classes.

Proof: We approach indirectly by assuming that it is. Then we have

$$\begin{aligned} 1 + 2 + \cdots + n &\equiv (a_1 + b_1) + (a_2 + b_2) + \cdots + (a_m + b_m) \\ &\equiv (a_1 + a_2 + \cdots + a_m) + (b_1 + b_2 + \cdots + b_m) \\ &\equiv 2(1 + 2 + \cdots + m) \pmod{m}, \end{aligned}$$

implying that $1 + 2 + \cdots + m \equiv 0 \pmod{m}$, or $m \mid \frac{m(m+1)}{2}$, which is not true for even integers m. Hence our assumption was wrong. □

Example 1.27. Let a be a positive integer. Determine all the positive integers m such that

$$\{a \cdot 1, a \cdot 2, a \cdot 3, \ldots, a \cdot m\}$$

is a set of complete residue classes modulo m.

Solution: The answer is the set of positive integers m that are relatively prime to a.

Let S_m denote the given set. First we show that S_m is a complete set of residue classes if $\gcd(a, m) = 1$. Because this set has exactly m elements, it suffices to show that elements in the set are not congruent to each other modulo m. Assume to the contrary that $ai \equiv aj \pmod{m}$ for some $1 \le i < j \le n$. Because $\gcd(a, m) = 1$, by Corollary 1.20, we have $i \equiv j \pmod{m}$, which is impossible since $|i - j| < m$. Hence our assumption was wrong and S_m is a complete set of residue classes modulo m.

On the other hand, if $g = \gcd(a, m) > 1$, then $a = a_1 g$ and $m = m_1 g$, where m_1 is a positive integer less than n. We have $am_1 \equiv a_1 m_1 g \equiv a_1 m \equiv am \equiv 0 \pmod{m}$. Hence S_m is not a complete set of residue classes. □

Similarly, we can show the following result.

Proposition 1.24. Let m be a positive integer. Let a be an integer relatively prime to m, and let b be an integer. Assume that S is a complete set of residue classes modulo m. The set

$$T = aS + b = \{as + b \mid s \in S\}$$

is also a complete set of residue classes modulo n.

Now we are better equipped to discuss linear congruence equations a bit further.

Proposition 1.25. Let m be a positive integer. Let a be an integer relatively prime to m, and let b be an integer. There exist integers x such that $ax \equiv b \pmod{m}$, and all these integers form exactly one residue class modulo m.

Proof: Let $\{c_1, c_2, \ldots, c_m\}$ be a complete set of residue classes modulo m. By Proposition 1.24,

$$\{ac_1 - b, ac_2 - b, \ldots, ac_m - b\}$$

is also a complete set of residue classes. Hence there exists c_i such that $ac_1 - b \equiv 0 \pmod{m}$, or c_1 is a solution to the congruence equation $ax \equiv b \pmod{m}$. It is easy to see that all the numbers congruent to c_1 modulo m also satisfy the congruence equation. On the other hand, if both x and x' satisfy the equation, we have $ax \equiv ax' \pmod{m}$. By Corollary 1.20, we have $x \equiv x' \pmod{m}$. □

In particular, setting $b = 1$ in Proposition 1.25 shows that if $\gcd(a, m) = 1$, then there is x such that $ax \equiv 1 \pmod{m}$. We call such x the **inverse of** a **modulo** m, denoted by a^{-1} or $\frac{1}{a} \pmod{m}$. Because all such numbers form exactly one residue class modulo m, the inverse of a is uniquely determined (or well defined) modulo m for all integers relatively prime to m.

Now we are ready to prove Wilson's theorem.

Theorem 1.26. [Wilson's Theorem] For any prime p, $(p-1)! \equiv -1 \pmod{p}$.

Proof: The property holds for $p = 2$ and $p = 3$, so we may assume that $p \geq 5$. Let $S = \{2, 3, \ldots, p-2\}$. Because p is prime, for any s in S, s has a unique inverse $s' \in \{1, 2, \ldots, p-1\}$. Moreover, $s' \neq 1$ and $s' \neq p-1$; hence $s' \in S$. In addition, $s' \neq s$; otherwise, $s^2 \equiv 1 \pmod{p}$, implying $p \mid (s-1)$ or $p \mid (s+1)$, which is not possible, since $s + 1 < p$. It follows that we can group the elements of S in $\frac{p-3}{2}$ distinct pairs (s, s') such that $ss' \equiv 1 \pmod{p}$. Multiplying these congruences gives $(p-2)! \equiv 1 \pmod{p}$ and the conclusion follows. □

Note that the converse of Wilson's theorem is true, that is, if $(n-1)! \equiv -1 \pmod{n}$ for an integer $n \geq 2$, then n is a prime. Indeed, if n were equal to $n_1 n_2$ for some integers $n_1, n_2 \geq 2$, we would have $n_1 \mid 1 \cdot 2 \cdots n_1 \cdots (n-1) + 1$, which is not possible. This provides us a new way to determine whether a number is prime. (However, this is not a very practical way, since for large n, $(n-1)!$ is huge!)

In most situations, there are no major differences in picking a particular complete set of residue classes to solve a particular problem. Here is a distinct example.

Example 1.28. [MOSP 2005, Melanie Wood] At each corner of a cube, an integer is written. A *legal transition* of the cube consists in picking any corner of the cube and adding the value written at that corner to the value written at some adjacent corner (that is, pick a corner with some value x written at it, and an adjacent corner with some value y written at it, and replace y by $x + y$). Prove that there is a finite sequence of legal transitions of the given cube such that the eight integers written are all the same modulo 2005.

We present two solutions. Notice that if we take a legal transition and perform it 2004 times, then modulo 2005, this is the same as replacing y by $y - x$. Call such a repetition of a single legal transition 2004 times a *super transition.*

First Solution: Look at the integers modulo 2005, and replace them with residue classes $1, 2, \ldots, 2005$. If all of the residue classes are the same, then we need no transitions. Otherwise, there is an edge with residue classes N and M with $1 \leq N < M \leq 2005$. Performing a super transition, we can replace M by $M - N$, which is a residue class, since $1 \leq M - N \leq 2005$. Since $N \geq 1$, this reduces the sum of the residue classes by at least 1. Because the sum of the residue classes is always at least 8, by repeating this process, we will eventually get to a state in which all of the residue classes are the same. □

Note that the proof would not work well if we replaced the numbers with residue classes $0, 1, \ldots, 2004$. As in the case $N = 0$, the sum of the residue classes is not decreased.

Second Solution: Look at the integers all modulo 2005. They are congruent to some set of positive integers modulo 2005. Performing a super transition on an edge is the same (modulo 2005) as performing a step of the Euclidean algorithm on the two numbers of the edge. Performing the Euclidean algorithm on a pair of positive integers will make them equal to the greatest common divisor of the two original integers after a finite number of steps. Thus, we can make two numbers of an edge congruent modulo 2005 after a finite number of super transitions. First we do this on all edges going in one direction, then on all the edges going in another direction, and then on all the edges going in the third direction. After this, we see that all the integers written at corners are congruent modulo 2005. □

Fermat's Little Theorem and Euler's Theorem

From the last few results, we note that for a given positive integer m, it is useful to consider the congruence classes that are relatively prime to m. For any positive integer m we denote by $\varphi(m)$ the number of all positive integers n less than m that are relatively prime to m. The function φ is called **Euler's totient function**. It is clear that $\varphi(1) = 1$ and for any prime p, $\varphi(p) = p - 1$. Moreover, if n is a positive integer such that $\varphi(n) = n - 1$, then n is a prime.

A set S of integers is also called a **reduced complete set of residue classes** modulo m if for each i with $0 \le i \le n-1$ and $\gcd(i, m) = 1$, there is an element $s \in S$ such that $i \equiv s \pmod{m}$. It is clear that a reduced complete set of residue classes modulo m consists of $\varphi(m)$ elements.

Proposition 1.27. Let m be a positive integer. Let a be an integer relatively prime to m. Assume that S is a reduced complete set of residue classes modulo m. Set

$$T = aS = \{as \mid s \in S\},$$

which is also a reduced complete set of residue classes modulo n.

The proof is similar to that of Proposition 1.24, and we leave it to the reader. Proposition 1.27 allows us to establish two of the most famous theorems in number theory.

Theorem 1.28. [Euler's Theorem] Let a and m be relatively prime positive integers. Then $a^{\varphi(m)} \equiv 1 \pmod{m}$.

Proof: Consider the set $S = \{a_1, a_2, \ldots, a_{\varphi(m)}\}$ consisting of all positive integers less than m that are relatively prime to m. Because $\gcd(a, n) = 1$, it follows from Proposition 1.26 that

$$\{aa_1, aa_2, \ldots, aa_{\varphi(m)}\}$$

is another reduced complete set of residue classes modulo n. Then

$$(aa_1)(aa_2)\cdots(aa_{\varphi(n)}) \equiv a_1 a_2 \cdots a_{\varphi(n)} \pmod{m}.$$

Using that $\gcd(a_k, n) = 1$, $k = 1, 2, \ldots, \varphi(n)$, the conclusion now follows. □

Setting $m = p$ as prime, Euler's theorem becomes Fermat's little theorem.

Theorem 1.29. [Fermat's Little Theorem] Let a be a positive integer and let p be a prime. Then

$$a^p \equiv a \pmod{p}.$$

Proof: We present an alternative proof independent of Euler's theorem. We induct on a. For $a = 1$ everything is clear. Assume that $p \mid (a^p - a)$. Then

$$(a+1)^p - (a+1) = (a^p - a) + \sum_{k=1}^{p-1} \binom{p}{k} a^k.$$

Using the fact that $p \mid \binom{p}{k}$ for $1 \le k \le p-1$ (Corollary 1.10) and the inductive hypothesis, it follows that p divides $(a+1)^p - (a+1)$; that is, $(a+1)^p \equiv (a+1) \pmod{p}$. □

Clearly, Fermat's little theorem is a special case of Euler's theorem. But with a few more properties on the Euler function φ that we will develop, we can derive Euler's theorem from Fermat's little theorem Note also another form of Fermat's little theorem:

Let a be a positive integer relatively prime to prime p. Then

$$a^{p-1} \equiv 1 \pmod{p}.$$

Next, we present a few examples involving these two important theorems.

Example 1.29. Let p be a prime. Prove that p divides $ab^p - ba^p$ for all integers a and b.

Proof: Note that $ab^p - ba^p = ab(b^{p-1} - a^{p-1})$.

If $p \mid ab$, then $p \mid ab^p - ba^p$; if $p \nmid ab$, then $\gcd(p, a) = \gcd(p, b) = 1$, and so $b^{p-1} \equiv a^{p-1} \equiv 1 \pmod{p}$, by Fermat's little theorem. Hence $p \mid b^{p-1} - a^{p-1}$, implying that $p \mid ab^p - ba^p$. Therefore, $p \mid ab^p - ba^p$ for all p. □

Example 1.30. Let $p \geq 7$ be a prime. Prove that the number

$$\underbrace{11\ldots1}_{p-1 \text{ 1's}}$$

is divisible by p.

Proof: We have

$$\underbrace{11\ldots1}_{p-1 \text{ 1's}} = \frac{10^{p-1} - 1}{9},$$

and the conclusion follows from Fermat's little theorem. (Note also that $\gcd(10, p) = 1$.) □

Example 1.31. Let p be a prime with $p > 5$. Prove that $p^8 \equiv 1 \pmod{240}$.

Proof: Note that $240 = 2^4 \cdot 3 \cdot 5$. By Fermat's little theorem, we have $p^2 \equiv 1 \pmod{3}$ and $p^4 \equiv 1 \pmod{5}$. Because a positive integer is relatively prime to 2^4 if and only if it is odd, $\varphi(2^4) = 2^3$. By Euler's theorem, we have $p^8 \equiv 1 \pmod{16}$. Therefore, $p^8 \equiv 1 \pmod{m}$ for $m = 3, 5$, and 16, implying that $p^8 \equiv 1 \pmod{240}$. □

Note that this solution indicates that we can establish Euler's theorem by Fermat's little theorem. Further, it is not difficult to check that $n^4 \equiv 1 \pmod{16}$ for $n \equiv \pm1, \pm3, \pm5, \pm7 \pmod{16}$ (see Example 1.24 (5)). Hence we can improve the result to $p^4 \equiv 1 \pmod{240}$ for all primes $p > 5$.

Example 1.32. Prove that for any even positive integer n, n^2-1 divides $2^{n!}-1$.

Proof: Let $m = n+1$. We need to prove that $m(m-2)$ divides $2^{(m-1)!}-1$. Because $\varphi(m)$ divides $(m-1)!$ we have $(2^{\varphi(m)}-1) \mid (2^{(m-1)!}-1)$ and from Euler's theorem, $m \mid (2^{\varphi(m)}-1)$. It follows that $m \mid (2^{(m-1)!}-1)$. Similarly, $(m-2) \mid (2^{(m-1)!}-1)$. Because m is odd, $\gcd(m, m-2) = 1$ and the conclusion follows. □

For a given positive integer m, let $\{a_1, a_2, \ldots, a_{\varphi(m)}\}$ be a reduced complete set of residue classes modulo m. By the existence and uniqueness of inverses, it is not difficult to see that the set of their inverses, denoted by

$$\{a_1^{-1}, a_2^{-1}, \ldots, a_{\varphi(m)}^{-1}\} \quad \text{or} \quad \left\{\frac{1}{a_1}, \frac{1}{a_2}, \ldots, \frac{1}{a_{\varphi(m)}}\right\},$$

is also a reduced complete set of residue classes modulo m. One might attempt to generalize Wilson's theorem by pairing residue classes that are inverses of each other. This approach fails, since there are residue classes other than 1 and -1 (or $m-1$) that are inverses of themselves. (In the proof of Wilson's theorem, there are only two possible values for s, namely $s = 1$ or $s = p-1$, such that $s^2 \equiv 1 \pmod{p}$.) For example, $6^2 \equiv 1 \pmod{35}$ for $m = 35$.

Let m be a positive integer, and let a be an integer relatively prime to m. Assume that $b = na$ is a multiple of a; that is, $n = \frac{b}{a}$ is an integer. From $a^{-1}a \equiv 1 \pmod{p}$, we have $n \equiv a^{-1}an \equiv a^{-1}b \pmod{m}$. This means that $n = \frac{a}{b}$ under the usual arithmetic meaning identifies with $n \equiv \frac{1}{a} \cdot b \pmod{m}$. This allows us to choose the order of operations to our advantage.

Example 1.33. [IMO 2005] Consider the sequence $a_1, a_2, \ldots$ defined by

$$a_n = 2^n + 3^n + 6^n - 1$$

for all positive integers n. Determine all positive integers that are relatively prime to every term of the sequence.

First Solution: The answer is 1. It suffices to show that every prime p divides a_n for some positive integer n. Note that both $p = 2$ and $p = 3$ divide $a_2 = 2^2 + 3^2 + 6^2 - 1 = 48$.

Assume now that $p \geq 5$. By Fermat's little theorem, we have $2^{p-1} \equiv 3^{p-1} \equiv 6^{p-1} \equiv 1 \pmod{p}$. Then

$$3 \cdot 2^{p-1} + 2 \cdot 3^{p-1} + 6^{p-1} \equiv 3 + 2 + 1 \equiv 6 \pmod{6},$$

or $6(2^{p-2} + 3^{p-2} + 6^{p-2} - 1) \equiv 0 \pmod{p}$; that is, $6a_{p-2}$ is divisible by p. Because p is relatively prime to 6, a_{p-2} is divisible by p, as desired.

Second Solution: If we use the notation of inverse, the proof can be written as

$$\begin{aligned} 6a_{p-2} &\equiv 6(2^{p-2} + 3^{p-2} + 6^{p-2} - 1) \\ &\equiv 6\left(\frac{1}{2} + \frac{1}{3} + \frac{1}{6} - 1\right) \equiv 0 \pmod{p}, \end{aligned}$$

for every prime p greater than 5. □

Example 1.34. Find an infinite nonconstant arithmetic progression of positive integers such that each term is not a sum of two perfect cubes.

Solution: Assume that the desired arithmetic progression is $\{a, a+d, a+2d, \dots\}$. We are basically considering all integers in the residue class a modulo d. We want to limit the number of residue classes that are cubes modulo d. In this way, we limit the number of residue classes that can be written as the sum of cubes modulo d.

We first look for d such that $a^3 \equiv 1 \pmod{d}$ for all integers a. Fermat's little theorem states that $a^{p-1} \equiv 1 \pmod{p}$ for prime p and integers a relatively prime to p. If we set $p - 1 = 3$, we have $p = 4$, which is not a prime. Hence we cannot apply Fermat's little theorem directly. On the other hand, if we set $p = 7$, then $a^6 \equiv 1 \pmod{7}$ for integers relatively prime to 7. It not difficult to check that the possible residue classes for a^3 modulo 7 are $0, 1, -1$ (or 6). Hence, modulo 7, the possible residue classes for $a^3 + b^3$ are $0, 1, -1, 2, -2$.

Therefore, $\{3, 3+7, 3+2\cdot 7, \dots\}$ and $\{4, 4+7, 4+2\cdot 7, \dots\}$ are two sequences satisfying the conditions of the problem. □

By noting that $\varphi(9) = 6$, we can also find sequences of the form $\{a, a+9, a+2\cdot 9, \dots\}$. We leave the details to the reader. (Compare to Example 1.24 (4).)

Example 1.35. [IMO 2003 shortlist] Determine the smallest positive integer k such that there exist integers $x_1, x_2, \dots, x_k$ with

$$x_1^3 + x_2^3 + \cdots + x_k^3 = 2002^{2002}.$$

Solution: The answer is $k = 4$.

We first show that 2002^{2002} is not a sum of three cubes. To restrict the number of cubes modulo n, we would like to have $\varphi(n)$ to be a multiple of 3. Again, consider $n = 7$. But adding three cubes modulo 7 gives too many residue classes (since 7 is too small). We then consider $n = 9$ with $\varphi(9) = 6$. Because $2002 \equiv 4 \pmod{9}$ and $2002^3 \equiv 4^3 \equiv 1 \pmod{9}$, it follows that

$$2002^{2002} \equiv (2002^3)^{667} \cdot 2004 \equiv 4 \pmod{9}.$$

On the other hand, $x^3 \equiv 0, \pm 1 \pmod{9}$ for integers x. We see that $x_1^3 + x_2^3 + x_3^3 \not\equiv 4 \pmod{9}$.

It remains to show that 2002^{2002} is a sum of four cubes. Starting with

$$2002 = 10^3 + 10^3 + 1^3 + 1^3$$

and using $2002 = 667 \cdot 3 + 1$ once again, we find that

$$\begin{aligned} 2002^{2002} &= 2002 \cdot (2002^{667})^3 \\ &= (10 \cdot 2002^{667})^3 + (10 \cdot 2002^{667})^3 + (2002^{667})^3 + (2002^{667})^3. \quad \square \end{aligned}$$

Fermat's little theorem provides a good criterion to determine whether a number is composite. But the converse is not true. For example, $3 \cdot 11 \cdot 17$ divides $a^{3\cdot 11\cdot 17} - a$, since 3, 11, 17 each divide $a^{3\cdot 11\cdot 17} - a$ (for instance, if 11 did not divide a, then from Fermat's little theorem, we have $11 \mid (a^{10} - 1)$; hence $11 \mid (a^{10\cdot 56} - 1)$, i.e., $11 \mid (a^{561} - a)$ and $561 = 3 \cdot 11 \cdot 17$).

The composite integers n satisfying $a^n \equiv a \pmod{n}$ for any integer a are called **Carmichael numbers**. There are also even Carmichael numbers, for example $n = 2 \cdot 73 \cdot 1103$.

Let a and m be relatively prime positive integers. Setting $b = 1$ in Corollary 1.23 leads to an interesting result. By Euler's theorem, there exist positive integers x such that $a^x \equiv 1 \pmod{m}$. We say that a has **order** d modulo m, denoted by $\text{ord}_m(a) = d$, if d is the smallest positive integer such that $a^d \equiv 1 \pmod{m}$. By Euler's theorem, $\text{ord}_m(a) = d \le \varphi(m)$. If x is a positive integer such that $a^x \equiv 1 \pmod{m}$, then by Corollary 1.23,

$$a^{\gcd(x,d)} \equiv 1 \pmod{m}.$$

Since $\gcd(x, d) \le d$, by the minimality of d, we must have $\gcd(x, d) = d$. Hence d divides x. We have established the following property.

Proposition 1.30. A positive integer x is such that $a^x \equiv 1 \pmod{m}$ if and only if x is a multiple of the order of a modulo m.

For a pair of relatively prime positive integers a and m, it is not true that there always exists a positive integer s such that $a^s \equiv -1 \pmod{n}$. (For example, $a = 2$ and $m = 7$.) Assume that there exists a perfect power of a that is congruent to -1 modulo m, and assume that s is the least such integer. We have $\text{ord}_m(a) = 2s$. Indeed, $a^{2s} \equiv 1 \pmod{m}$, so d divides $2s$. If $d < 2s$, then $2^{2s-d} \equiv -1 \pmod{m}$ violates the minimality assumption on s. Furthermore, if t is an integer such that

$$a^t \equiv -1 \pmod{m},$$

then t is a multiple of s. Because $a^{2t} \equiv 1 \pmod{m}$, it follows that $d = 2s$ divides $2t$, and so s divides t. It is then clear that t must an odd multiple of s; that is,

$$a^t \equiv \begin{cases} -1 & \text{if } t \text{ is an odd mulptiple of } s; \\ 1 & \text{if } t \text{ is an even mulptiple of } s. \end{cases}$$

Example 1.36. [AIME 2001] How many positive integer multiples of 1001 can be expressed in the form $10^j - 10^i$, where i and j are integers and $0 \le i < j \le 99$?

Solution: Because

$$10^j - 10^i = 10^i(10^{j-i} - 1)$$

and $1001 = 7 \cdot 11 \cdot 13$ is relatively prime to 10^i, it is necessary to find i and j such that the $10^{j-i} - 1$ is divisible by the primes 7, 11, and 13. Notice that $1001 = 7 \cdot 11 \cdot 13$; that is, $10^3 \equiv -1 \pmod{1001}$. It is easy to check that

$$\text{ord}_{1001}(10) = 6.$$

By Proposition 1.30, $10^i(10^{j-i} - 1)$ is divisible by 1001 if and only if $j - i = 6n$ for some positive integer n. Thus it is necessary to count the number of integer solutions to

$$i + 6n = j,$$

where $j \leq 99$, $i \geq 0$, and $n > 0$. For each $n = 1, 2, 3, \ldots, 15$, there are $100 - 6n$ suitable values of i (and j), so the number of solutions is

$$94 + 88 + 82 + \cdots + 4 = 784.$$

□

Euler's Totient Function

We discuss some useful properties of Euler's totient function φ. First of all, it is not difficult to see the following:

Proposition 1.31. Let p be a prime, and let a be a positive integer. Then $\varphi(p^a) = p^a - p^{a-1}$.

Next, we show that φ is multiplicative:

Proposition 1.32. Let a and b be two relatively prime positive integers. Then $\varphi(ab) = \varphi(a)\varphi(b)$.

Proof: : Arrange the integers $1, 2, \ldots, ab$ into an $a \times b$ array as follows:

$$\begin{array}{cccc} 1 & 2 & \cdots & a \\ a+1 & a+2 & \cdots & 2a \\ \vdots & \vdots & \vdots & \vdots \\ a(b-1)+1 & a(b-1)+2 & \cdots & ab. \end{array}$$

Clearly, there are $\varphi(ab)$ numbers in the above table that are relatively prime to ab.

On the other hand, there are $\varphi(a)$ columns containing those elements in the table relatively prime to a. Each of those columns is a complete set of residues modulo b, by Proposition 1.24. Hence there are exactly $\varphi(b)$ elements in each of those columns that are relatively prime to b. Therefore, there are $\varphi(a)\varphi(b)$ numbers in the table that are relatively prime to ab.

Hence $\varphi(ab) = \varphi(a)\varphi(b)$ for relatively prime integers ab. □

Theorem 1.33. If $n = p_1^{\alpha_1} \cdots p_k^{\alpha_k}$ is the prime factorization of $n > 1$, then

$$\varphi(n) = n\left(1 - \frac{1}{p_1}\right)\cdots\left(1 - \frac{1}{p_k}\right).$$

First Proof: This follows directly from Propositions 1.31 and 1.32. □

Second Proof: We employ the inclusion and exclusion principle. Set

$$T_i = \{d \,:\, d \le n \text{ and } p_i | d\},$$

for $i = 1, \ldots, k$. It follows that

$$T_1 \cup \cdots \cup T_k = \{m \,:\, m \le n \text{ and } \gcd(m, n) > 1\}.$$

Hence

$$\begin{aligned}\varphi(n) &= n - |T_1 \cup \cdots \cup T_k| \\ &= n - \sum_{i=1}^{k} |T_i| + \sum_{1 \le i < j \le k} |T_i \cap T_j| - \cdots + (-1)^k |T_1 \cap \cdots \cap T_k|.\end{aligned}$$

We have

$$|T_i| = \frac{n}{p_i}, \quad |T_i \cap T_j| = \frac{n}{p_i p_j}, \quad \ldots, \quad |T_1 \cap \cdots \cap T_k| = \frac{n}{p_1 \cdots p_k}.$$

Finally,

$$\begin{aligned}\varphi(n) &= n\left(1 - \sum_{i=1}^{n} \frac{1}{p_i} + \sum_{1 \le i < j \le k} \frac{1}{p_i p_j} - \cdots + (-1)^k \frac{1}{p_1 \cdots p_k}\right) \\ &= n\left(1 - \frac{1}{p_1}\right)\cdots\left(1 - \frac{1}{p_k}\right).\end{aligned}$$ □

Based on Theorem 1.33, we can establish Euler's theorem from Fermat's little theorem. Indeed, let $n = p_1^{\alpha_1} \cdots p_k^{\alpha_k}$ be the prime factorization of n. We have $a^{p_i - 1} \equiv 1 \pmod{p_i}$, hence $a^{p_i(p_i-1)} \equiv 1 \pmod{p_i^2}$, $a^{p_i^2(p_i-1)} \equiv 1 \pmod{p_i^3}, \ldots, a^{p_i^{\alpha_i-1}(p_i-1)} \equiv 1 \pmod{p_i^{\alpha_i}}$. That is, $a^{\varphi(p_i^{\alpha_i})} \equiv 1 \pmod{p_i^{\alpha_i}}$, $i = 1, \ldots, k$. Applying this property to each prime factor, the conclusion follows.

Theorem 1.34. [Gauss] For any positive integer n,

$$\sum_{d|n} \varphi(d) = n.$$

Proof: We consider the rational numbers

$$\frac{1}{n}, \frac{2}{n}, \ldots, \frac{n}{n}.$$

Clearly, there are n numbers in the list.

We obtain a new list by reducing each number in the above list to the lowest terms; that is, express each fraction as a quotient of relatively prime integers. The denominators of the numbers in the new list will all be divisors of n. If $d \mid n$, exactly $\varphi(d)$ of the numbers in the list will have d as their denominator. (This is the meaning of lowest terms!) Hence, there are $\sum_{d|n} \varphi(d)$ in the new list.

Because the two lists have the same number of terms, we obtain the desired result. □

Example 1.37. Let n be a positive integer.

(1) Find the sum of all positive integers less than n and relatively prime to n.

(2) Find the sum of all positive integers less than $2n$ and relatively prime to n.

Solution: The answers are $\frac{n\varphi(n)}{2}$ and $2n\varphi(n)$, respectively.

Let

$$S_1 = \sum_{\substack{d<n \\ \gcd(d,n)=1}} d \quad \text{and} \quad S_2 = \sum_{\substack{d<2n \\ \gcd(d,n)=1}} d.$$

Let $d_1 < d_2 < \cdots < d_{\varphi(n)}$ be the numbers less than n and relatively prime to n. Note that $\gcd(d, n) = 1$ if and only if $\gcd(n - d, n) = 1$. We deduce that

$$d_1 + d_{\varphi(n)} = n, \quad d_2 + d_{\varphi(n)-1} = n, \quad \ldots, \quad d_{\varphi(n)} + d_1 = n,$$

implying that

$$S_1 = \frac{n\varphi(n)}{2}.$$

On the other hand,

$$\sum_{\substack{n<d<2n \\ \gcd(d,n)=1}} d = \sum_{\substack{d<n \\ \gcd(d,n)=1}} (n+d) = n\varphi(n) + \sum_{\substack{d<n \\ \gcd(d,n)=1}} d$$

$$= n\varphi(n) + \frac{n\varphi(n)}{2} = \frac{3n\varphi(n)}{2}.$$

Therefore

$$S_2 = \frac{n\varphi(n)}{2} + \frac{3n\varphi(n)}{2} = 2n\varphi(n).$$

□

Multiplicative Function

We include this section to further develop results related to three functions we already introduced: $\tau(n)$ (the number of the positive divisors of n), $\sigma(n)$ (the sum of the positive divisors of n), and $\varphi(n)$ (Euler's totient function). This might well be the most abstract part of this book, and the material covered in this section is not essential to the rest of the book. However, it is very useful for further study in number theory.

Arithmetic functions are defined on the positive integers and are complex valued. The arithmetic function $f \neq 0$ is called **multiplicative** if for any relatively prime positive integers m and n,

$$f(mn) = f(m)f(n).$$

Note that if f is multiplicative, then $f(1) = 1$. Indeed, if a is a positive integer for which $f(a) \neq 0$, then $f(a) = f(a \cdot 1) = f(a)f(1)$, and simplifying by $f(a)$ yields $f(1) = 1$. Note also that if f is multiplicative and $n = p_1^{\alpha_1} \cdots p_k^{\alpha_k}$ is the prime factorization of the positive integer n, then $f(n) = f(p_1^{\alpha_1}) \cdots f(p_k^{\alpha_k})$.

An important arithmetic function is the **Möbius function** defined by

$$\mu(n) = \begin{cases} 1 & \text{if } n = 1, \\ 0 & \text{if } p^2 \mid n \text{ for some prime } p > 1, \\ (-1)^k & \text{if } n = p_1 \cdots p_k, \text{ where } p_1, \ldots, p_k \text{ are distinct primes.} \end{cases}$$

For example, $\mu(2) = -1$, $\mu(6) = 1$, $\mu(12) = \mu(2^2 \cdot 3) = 0$.

Theorem 1.35. The Möbius function μ is multiplicative.

Proof: Let m, n be positive integers such that $\gcd(m, n) = 1$. If $p^2 \mid m$ for some $p > 1$, then $p^2 \mid mn$ and so $\mu(m) = \mu(mn) = 0$ and we are done. Consider now $m = p_1 \cdots p_k$, $n = q_1 \cdots q_h$, where $p_1, \ldots, p_k, q_1, \ldots, q_h$ are distinct primes. Then $\mu(m) = (-1)^k$, $\mu(n) = (-1)^h$, and $mn = p_1 \cdots p_k q_1 \cdots q_h$. It follows that $\mu(mn) = (-1)^{k+h} = (-1)^k(-1)^h = \mu(m)\mu(n)$. □

For an arithmetic function f we define its **summation function** F by

$$F(n) = \sum_{d|n} f(d).$$

The connection between f and F is given by the following result.

Theorem 1.36. If f is multiplicative, then so is its summation function F.

Proof: Let m and n be positive relatively prime integers and let d be a divisor of mn. Then d can be uniquely represented as $d = kh$, where $k \mid m$ and $h \mid n$.

Because $\gcd(m, n) = 1$, we have $\gcd(k, h) = 1$, so $f(kh) = f(k)f(h)$. Hence

$$F(mn) = \sum_{d|mn} f(d) = \sum_{\substack{k|m \\ h|n}} f(k)f(h)$$
$$= \left(\sum_{k|m} f(k)\right)\left(\sum_{h|h} f(h)\right) = F(m)F(n). \qquad \square$$

Note that if f is a multiplicative function and $n = p_1^{\alpha_1} \cdots p_k^{\alpha_k}$, then

$$\sum_{d|n} \mu(d)f(d) = (1 - f(p_1)) \cdots (1 - f(p_k)).$$

Indeed, the function $g(n) = \mu(n)f(n)$ is multiplicative; hence from Theorem 1.36, so is its summation function G. Then $G(n) = G(p_1^{\alpha_1}) \cdots G(p_k^{\alpha_k})$ and

$$G(p_i^{\alpha_i}) = \sum_{d|p_i^{\alpha_i}} \mu(d)f(d) = \mu(1)f(1) + \mu(p_i)f(p_i) = 1 - f(p_i),$$

and the conclusion follows.

Theorem 1.37. [Möbius inversion formula] Let f be an arithmetic function and let F be its summation function. Then

$$f(n) = \sum_{d|n} \mu(d)F\left(\frac{n}{d}\right).$$

Proof: We have

$$\sum_{d|n} \mu(d)F\left(\frac{n}{d}\right) = \sum_{d|n} \mu(d)\left(\sum_{c|\frac{n}{d}} f(c)\right) = \sum_{d|n}\left(\sum_{c|\frac{n}{d}} \mu(d)f(c)\right)$$
$$= \sum_{c|n}\left(\sum_{d|\frac{n}{c}} \mu(d)f(c)\right) = \sum_{c|n} f(c)\left(\sum_{d|\frac{n}{c}} \mu(d)\right) = f(n),$$

since for $\frac{n}{c} > 1$ we have $\sum_{d|\frac{n}{c}} \mu(d) = 0$.

We have used the fact that

$$\left\{(d, c)|\ d|n \text{ and } c|\frac{n}{d}\right\} = \left\{(d, c)|\ c|n \text{ and } d|\frac{n}{c}\right\}. \qquad \square$$

Theorem 1.38. Let f be an arithmetic function and let F be its summation function. If F is multiplicative, then so is f.

Proof: Let m, n be positive integers such that $\gcd(m, n) = 1$ and let d be a divisor of mn. Then $d = kh$, where $k \mid m$, $h \mid n$, and $\gcd(k, h) = 1$. Applying the Möbius inversion formula it follows

$$\begin{aligned} F(mn) &= \sum_{d \mid mn} \mu(d) F\left(\frac{mn}{d}\right) = \sum_{\substack{k \mid m \\ h \mid n}} \mu(kh) F\left(\frac{mn}{kh}\right) \\ &= \sum_{\substack{k \mid m \\ h \mid n}} \mu(k)\mu(h) F\left(\frac{m}{k}\right) F\left(\frac{n}{h}\right) \\ &= \left(\sum_{k \mid m} \mu(k) F\left(\frac{m}{k}\right)\right) \left(\sum_{h \mid n} \mu(h) F\left(\frac{n}{h}\right)\right) \\ &= f(m) f(n). \end{aligned}$$

□

We leave it to the reader to show that functions τ, σ, and φ are indeed multiplicative. We also encourage the reader to redevelop some properties of these functions by the general results we developed in this section.

Linear Diophantine Equations

An equation of the form

$$a_1 x_1 + \cdots + a_n x_n = b, \qquad (*)$$

where $a_1, a_2, \ldots, a_n, b$ are fixed integers, is called a **linear Diophantine equation**. We assume that $n \geq 1$ and that coefficients $a_1, \ldots, a_n$ are all different from zero.

The main result concerning linear Diophantine equations is the following generalization of Theorem 1.7. (Bézout's identity).

Theorem 1.39. The equation $(*)$ is solvable if and only if

$$\gcd(a_1, \ldots, a_n) \mid b.$$

In case of solvability, all integer solutions to $(*)$ can be expressed in terms of $n-1$ integral parameters.

Proof: Let $d = \gcd(a_1, \ldots, a_n)$.

If b is not divisible by d, then $(*)$ is not solvable, since for any integers $x_1, \ldots, x_n$ the left-hand side of $(*)$ is divisible by d and the right-hand side is not.

If $d \mid b$, then we obtain the equivalent equation

$$a_1' x_1 + \cdots + a_n' x_n = b',$$

where $a_i' = a_i/d$ for $i = 1, \ldots, n$ and $b' = b/d$. Clearly, we have $\gcd(a_1', \ldots, a_n') = 1$.

We use induction on the number n of the variables. In the case $n = 1$ the equation has the form $x_1 = b$ or $-x_1 = b$, and thus the unique solution does not depend on any parameter.

We now assume that $n \geq 2$ and that the solvability property holds for all linear equations in $n - 1$ variables. Our goal is to prove the solvability of equations in n variables. Set $d_{n-1} = \gcd(a_1, \ldots, a_{n-1})$. Then any solution $(x_1, \ldots, x_n)$ of (1) satisfies the congruence

$$a_1x_1 + a_2x_2 + \cdots + a_nx_n \equiv b \pmod{d_{n-1}},$$

which is equivalent to

$$a_nx_n \equiv b \pmod{d_{n-1}}. \tag{†}$$

Multiplying both sides of (†) by $a_n^{\varphi(d_{n-1})-1}$ and taking into account that $a_n^{\varphi(d_{n-1})} \equiv 1 \pmod{d_{n-1}}$, we obtain

$$x_n \equiv c \pmod{d_{n-1}},$$

where $c = a_n^{\varphi(d_{n-1})-1}b$. It follows that $x_n = c + d_{n-1}t_{n-1}$ for some integer t_{n-1}. Substituting in (∗) and rearranging yields the equation in $(n - 1)$ variables

$$a_1x_1 + \cdots + a_{n-1}x_{n-1} = b - a_nc - a_{n-1}d_{n-1}t_{n-1}.$$

It remains to show that $d_{n-1} \mid (b - a_nc - a_{n-1}d_{n-1}t_{n-1})$, which is equivalent to $a_nc \equiv b \pmod{d_{n-1}}$. The last relation is true because of the choice of c. Therefore we can divide the last equation by d_{n-1}, and obtain

$$a_1'x_1 + \cdots + a_{n-1}'x_{n-1} = b', \tag{‡}$$

where $a_i' = a_i/d_{n-1}$ for $i = 1, \ldots, n - 1$ and $b' = (b - a_nc)/d_{n-1} - a_nt_{n-1}$. Because $\gcd(a_1', \ldots, a_{n-1}') = 1$, by the induction hypothesis the equation (‡) is solvable for each integer t_{n-1} and its solutions can be written in terms of $n - 2$ integral parameters. If we add to these solutions $x_n = c + d_{n-1}t_{n-1}$, we obtain solutions to (∗) in terms of $n - 1$ parameters. □

Corollary 1.40. Let a_1, a_2 be relatively prime integers. If (x_1^0, x_2^0) is a solution to the equation

$$a_1x_1 + a_2x_2 = b,$$

then all of its solutions are given by

$$\begin{cases} x_1 = x_1^0 + a_2t, \\ x_2 = x_2^0 - a_1t, \end{cases}$$

for every integer t.

Example 1.38. Determine all triples (x, y, z) of integers satisfying the equation $3x + 4y + 5z = 6$.

Solution: We have $3x + 4y \equiv 1 \pmod 5$; hence

$$3x + 4y = 1 + 5s$$

for some integer s. A solution to this equation is $x = -1 + 3s$, $y = 1 - s$. Applying Corollary 1.40, we obtain $x = -1 + 3s + 4t$ and $y = 1 - s - 3t$, for some integer t, and substituting back into the original equation yields $z = 1 - s$. Hence all solutions are

$$(x, y, z) = (-1 + 3s + 4t, 1 - s - 3t, 1 - s),$$

for all pairs of integers s and t, □

Example 1.39. Let n be a positive integer. Suppose that there are 666 ordered triples (x, y, z) of positive integers satisfying the equation

$$x + 8y + 8z = n.$$

Find the maximum value of n.

Solution: The answer is 303.

Write $n = 8a + b$, where a and b are integers with $0 \le b < 8$. Since $x \equiv n \equiv b \pmod 8$, the possible values of x are $b, 8 + b, \ldots, 8(a - 1) + b$. For $x = b + 8i$, where $0 \le i \le a - 1$, $8(y + z) = 8(a - i)$ or $y + z = a - i$, which admits $a - i - 1$ ordered pairs (y, z) of positive integer solutions, namely, $(1, a - i - 1), \ldots, (a - i - 1, 1)$. Hence there are

$$\sum_{i=0}^{a-1}(a - i - 1) = \sum_{i=0}^{a-1} i = \frac{a(a - 1)}{2}$$

ordered triples satisfying the conditions of the problem. Solving $\frac{a(a-1)}{2} = 666$ gives $a = 37$. Therefore, the maximum value for n is equal to $37 \cdot 8 + 7 = 303$, obtained by setting $b = 7$. □

Numerical Systems

The fundamental result in this subsection is given by the following theorem:

Theorem 1.41. Let b be an integer greater than 1. For any integer $n \ge 1$ there is a unique system $(k, a_0, a_1, \ldots, a_k)$ of integers such that $0 \le a_i \le b - 1$, $i = 0, 1, \ldots, k$, $a_k \ne 0$, and

$$n = a_k b^k + a_{k-1} b^{k-1} + \cdots + a_1 b + a_0. \qquad (*)$$

Proof: For the existence, we apply repeatedly the division algorithm:

$$n = q_1 b + r_1, \quad 0 \le r_1 \le b-1;$$
$$q_1 = q_2 b + r_2, \quad 0 \le r_2 \le b-1;$$
$$\dots$$
$$q_{k-1} = q_k b + r_k, \quad 0 \le r_k \le b-1;$$

where q_k is the last nonzero quotient.

Let

$$q_0 = n,\ a_0 = n - q_1 b,\ a_1 = q_1 - q_2 b, \dots, a_{k-1} = q_{k-1} - q_k b,\ a_k = q_k.$$

Then

$$\sum_{i=0}^{k} a_i b^i = \sum_{i=0}^{k-1} (q_i - q_{i+1} b) b^i + q_k b^k = q_0 + \sum_{i=1}^{k} q_i b^i - \sum_{i=1}^{k} q_i b^i = q_0 = n.$$

For uniqueness, assume that $n = c_0 + c_1 b + \cdots + c_n b^n$ is another such representation.

If $h \neq k$, for example $h > k$, then $n \ge b^h \ge b^{k+1}$. But

$$n = a_0 + a_1 b + \cdots + a_k b^k \le (b-1)(1 + b + \cdots + b^k) = b^{k+1} - 1 < b^{k+1},$$

a contradiction.

If $h = k$, then

$$a_0 + a_1 b + \cdots + a_k b^k = c_0 + c_1 b + \cdots + c_k b^k,$$

and so $b \mid (a_0 - c_0)$. On the other hand, $|a_0 - c_0| < b$; hence $a_0 = c_0$, Therefore

$$a_1 + a_2 b + \cdots + a_k b^{k-1} = c_1 + c_2 b + \cdots + c_k b^{k-1}.$$

By repeating the above procedure, it follows that $a_1 = c_1,\ a_2 = c_2, \dots,$ and $a_k = c_k$. □

Relation (∗) is called *the base-b representation* of n and is denoted by

$$n = \overline{a_k a_{k-1} \dots a_0}_{(b)}.$$

The usual **decimal representation** corresponds to $b = 10$ and we write only $n = a_k a_{k-1} \dots a_0$ instead. (For example, $4567 = \overline{4567}_{(10)}$.)

Example 1.40. Let $\overline{xy}$ and $\overline{yx}$ be two 2-digit integers. Prove that their sum is composite.

Proof: Since $\overline{xy} = 10x + y$ and $\overline{yx} = 10y + x$, their sum is equal to $11x + 11y = 11(x + y)$, a composite number. □

Example 1.41. [AHSME 1973] In the following equation, each of the letters represents uniquely a different digit in base ten:

$$(YE) \cdot (ME) = TTT.$$

Determine the sum $E + M + T + Y$.

Solution: Because $TTT = T \cdot 111 = T \cdot 3 \cdot 37$, one of YE and ME is 37, implying that $E = 7$. But T is a digit and $T \cdot 3$ is a two-digit number ending with 7, and so it follows that $T = 9$ and $TTT = 999 = 27 \cdot 37$, and so $E+M+T+Y = 2+3+7+9 = 21$. □

Example 1.42. [AIME 2001] Find the sum of all positive two-digit integers that are divisible by each of their digits.

Solution: Let $\overline{ab}$ denote an integer with the required property. Then $10a + b$ must be divisible by both a and b. It follows that b must be divisible by a, and that $10a$ must be divisible by b. The former condition requires that $b = ka$ for some positive integer k, and the latter condition implies that $k = 1$ or $k = 2$ or $k = 5$. Thus the requested two-digit numbers are $11, 22, \ldots, 99, 12, 24, 36, 48$, and 15. Their sum is $11 \cdot 45 + 12 \cdot 10 + 15 = 630$. □

Example 1.43. [AMC12A 2002] Some sets of prime numbers, such as $\{7, 83, 421, 659\}$, use each of the nine nonzero digits exactly once. What is the smallest possible sum such a set of primes can have?

Solution: The answer is 207.

Note that digits 4, 6, and 8 cannot appear in the units digit. Hence the sum is at least $40 + 60 + 80 + 1 + 2 + 3 + 5 + 7 + 9 = 207$. On the other hand, this value can be obtained with the set $\{2, 5, 7, 43, 61, 89\}$. □

Example 1.44. Write $\overline{101011}_{(2)}$ in base 10, and write 1211 in base 3.

Solution: We have

$$\begin{aligned}\overline{1010011}_{(2)} &= 1 \cdot 2^6 + 0 \cdot 2^5 + 1 \cdot 2^4 + 0 \cdot 2^3 + 0 \cdot 2^2 + 1 \cdot 2 + 1 \\ &= 64 + 16 + 2 + 1 = 83.\end{aligned}$$

Dividing by 3 successively, the remainders give the digits of the base-3 representation, beginning with the last. The first digit is the last nonzero quotient. We can arrange the computations as follows:

```
1211| 3
1209 403| 3
   2  402 134| 3
        1 132 44| 3
            2  42 14|3
                2 12 4|3
                   2 3 1
                       1
```

Hence $1211 = \overline{1122212}_{(3)}$. □

Example 1.45. The product of seven and the six-digit number $\overline{abcdef}$ is equal to the product of six and the six-digit number $\overline{defabc}$. Find these two six-digit numbers.

Solution: Let x and y denote the three-digit numbers $\overline{abc}$ and $\overline{def}$, respectively. Then $\overline{abcdef} = 1000x+y$ and $\overline{defabc} = 1000y+x$. By the given conditions, we have $7(1000x+y) = 6(1000y+x)$, or $6994x = 5993y$. Since $\gcd(6994, 5993) = \gcd(5993, 1001) = \gcd(1001, 13) = 13$, we have $538x = 461y$, and so the two numbers are 461538 and 538461. □

Example 1.46. [AMC12A 2005] A faulty car odometer proceeds from digit 3 to digit 5, always skipping the digit 4, regardless of position. For example, after traveling one mile the odometer changed from 000039 to 000050. If the odometer now reads 002005, how many miles has the car actually traveled?

Solution: Because the odometer uses only 9 digits, it records mileage in base-9 numerals, except that its digits 5, 6, 7, 8, and 9 represent the base-9 digits 4, 5, 6, 7, and 8. Therefore the mileage is

$$2004_{(9)} = 2 \cdot 9^3 + 4 = 2 \cdot 729 + 4 = 1462.$$ □

Example 1.47. Prove that the number $11\ldots1_{(9)}$ in base 9 is triangular; that is, it is the sum of the first k positive integers for some positive integer k.

Proof: Indeed

$$\begin{aligned} \underbrace{11\ldots1}_{n\ 1\text{'s}}{}_{(9)} &= 9^{n-1} + 9^{n-2} + \cdots + 9 + 1 \\ &= \frac{9^n - 1}{9 - 1} = \frac{1}{2} \cdot \frac{3^n - 1}{2} \cdot \frac{3^n + 1}{2} \\ &= 1 + 2 + \cdots + \frac{3^n - 1}{2}. \end{aligned}$$

Thus it is a triangular number. □

Example 1.48. Determine all positive integers n such that $11111_{(n)}$ is a perfect square.

Solution: The answer is $n = 3$.

We have $11111_{(n)} = n^4 + n^3 + n^2 + n + 1$.

If n is even, then $n^2 + \frac{n}{2}$ and $n^2 + \frac{n}{2} + 1$ are two consecutive integers. We have

$$\begin{aligned}\left(n^2 + \frac{n}{2}\right)^2 &= n^4 + n^3 + \frac{n^2}{4} \\ &< n^4 + n^3 + n^2 + n + 1 \\ &< \left(n^2 + \frac{n}{2} + 1\right)^2.\end{aligned}$$

Hence $11111_{(n)}$ is not a perfect square for even positive integers n.

If n is odd, then $n^2 + \frac{n}{2} - \frac{1}{2}$ and $n^2 + \frac{n}{2} + \frac{1}{2}$ are integers. Clearly, we have

$$\left(n^2 + \frac{n}{2} - \frac{1}{2}\right)^2 < n^4 + n^3 + n^2 + n + 1.$$

Note that

$$\begin{aligned}\left(n^2 + \frac{n}{2} + \frac{1}{2}\right)^2 &= n^4 + n^3 + \frac{5n^2}{4} + \frac{n}{2} + \frac{1}{4} \\ &= n^4 + n^3 + n^2 + n + 1 + \frac{n^2 - 2n - 3}{4} \\ &= n^4 + n^3 + n^2 + n + 1 + \frac{(n-3)(n+1)}{4}.\end{aligned}$$

For odd integers n greater than 3, $11111_{(n)}$ is strictly between two consecutive perfect squares, namely,

$$\left(n^2 + \frac{n}{2} - \frac{1}{2}\right)^2 \quad \text{and} \quad \left(n^2 + \frac{n}{2} + \frac{1}{2}\right)^2.$$

Hence $11111_{(n)}$ is not a perfect square for any positive integers other than 3. For $n = 3$, we have $11111_{(3)} = 121 = 11^2$. □

In the last example, we showed that an integer is not a perfect square by placing the integer between two consecutive perfect squares. This method works because integers are discrete. Such methods will hardly work for real numbers, since there are no holes in between real numbers. This is a very useful method in solving Diophantine equations.

In certain numerical systems, the base does not have to be constant. Here are two examples.

Proposition 1.42. Every positive integer k has a unique **factorial base expansion**

$$(f_1, f_2, f_3, \dots, f_m),$$

meaning that

$$k = 1! \cdot f_1 + 2! \cdot f_2 + 3! \cdot f_3 + \cdots + m! \cdot f_m,$$

where each f_i is an integer, $0 \le f_i \le i$, and $f_m > 0$.

Proof: Note that there exists a unique positive integer m_1 such that $m_1! \le k < (m_1 + 1)!$. By the division algorithm, we can write

$$k = m_1! f_{m_1} + r_1$$

for some positive integer f_{m_1} and some integer r_1 with $0 \le r_1 < m_1!$. Because $k < (m + 1)! = m! \cdot (m + 1)$, it follows that $f_{m_1} \le m$. Repeating this process, we can then write

$$r_1 = m_2! f_{m_2} + r_2,$$

with m_2 the unique positive integer with $m_2! \le r_1 < (m_2 + 1)!$, $1 \le f_{m_2} \le m_2$, and $0 \le r_2 < m_2!$. Keeping this process on r_2, and so on, we obtain a unique factorial base expansion of k. □

Proposition 1.43. Let $F_0 = 1$, $F_1 = 1$, and $F_{n+1} = F_n + F_{n-1}$ for every positive integer n. (This sequence is called the **Fibonacci sequence**, and its terms are called **Fibonacci numbers**.) Each nonnegative integer n can be uniquely written as a sum of nonconsecutive positive Fibonacci numbers; that is, each nonnegative integer n can be written uniquely in the form

$$n = \sum_{k=0}^{\infty} \alpha_k F_k,$$

where $\alpha_k \in \{0, 1\}$ and $(\alpha_k, \alpha_{k+1}) \ne (1, 1)$ for each k. This expression for n is called its **Zeckendorf representation**.

The proof of Proposition 1.43 is similar to that of Proposition 1.42, and we leave the details to the reader.

Example 1.49. [AIME2 2000] Given that $(f_1, f_2, f_3, \dots, f_j)$ is the factorial base expansion of

$$16! - 32! + 48! - 64! + \cdots + 1968! - 1984! + 2000!,$$

find the value of $f_1 - f_2 + f_3 - f_4 + \cdots + (-1)^{j-1} f_j$.

Solution: Because $(n+1)! - n! = n!(n+1) - n! = n!n$, it follows that

$$\begin{aligned}(n+16)! - n! &\\ &= (n+16)! - (n+15)! + (n+15)! - (n+14)! + \cdots + (n+1)! - n! \\ &= (n+15)!(n+15) + (n+14)!(n+14) + \cdots + (n+1)!(n+1) + n!n.\end{aligned}$$

This shows that the factorial base expansion of $(n+16)! - n!$ is

$$(0, 0, \ldots, 0, n, n+1, \ldots, n+14, n+15),$$

which begins with a block of $n-1$ zeros. The factorial base expansion of $16!$ is $(0, 0, \ldots, 0, 1)$, so the requested expansion is

$$\begin{aligned}(0, 0, \ldots, 0; 1; 0, \ldots, 0; 32, 33, \ldots, 47; &\\ 0, \ldots, 0; 64, \ldots, 79; \ldots; 1984, \ldots, 1999).\end{aligned}$$

Notice that starting in position thirty-two, the expansion contains groups of sixteen nonzero numbers alternating with groups of sixteen zeros. With the exception of $f_{16} = 1$, each nonzero f_i is i. Each of the 62 groups of sixteen nonzero numbers contributes 8 to the alternating sum, and f_{16} contributes -1, so the requested value is $8 \cdot 62 - 1 = 495$. □

Divisibility Criteria in the Decimal System

We will prove some divisibility criteria for integers in decimal representation.

Proposition 1.44. Let $n = \overline{a_h a_{h-1} \ldots a_0}$ be a positive integer.

(a) Let $s(n) = a_0 + a_1 + \cdots + a_h$ denote the sum of its digits. Then $n \equiv s(n) \pmod 3$. In particular, n is divisible by 3 if and only if the sum $S(n)$ of its digits is divisible by 3.

(b) We can replace 3 by 9 in (1); that is, $n \equiv S(n) \pmod 9$. In particular, n is divisible by 9 if and only if the sum $S(n)$ of its digits is divisible by 9.

(c) Let $s'(n) = a_0 - a_1 + \cdots + (-1)^h a_h$ (*alternating sum*). Then n is divisible by 11 if and only if $s'(n)$ is divisible by 11.

(d) n is divisible by 7, 11, or 13 if and only if $\overline{a_h a_{h-1} \ldots a_3} - \overline{a_2 a_1 a_0}$ has this property.

(e) n is divisible by 27 or 37 if and only if $\overline{a_h a_{h-1} \ldots a_3} + \overline{a_2 a_1 a_0}$ has this property.

(f) n is divisible by 2^k or 5^k $(k \leq h)$ if and only if $\overline{a_{k-1} \dots a_0}$ has this property.

Proof: For (a) and (b), since $10^k = (9+1)^k$, it follows that $10^k \equiv 1 \pmod 9$. Hence $n \equiv \sum_{k=0}^{h} a_k 10^k = \sum_{k=0}^{h} a_k \equiv S(n) \pmod 9$.

For (c), we note that $10^k = (11-1)^k$. Hence $10^k \equiv (-1)^k \pmod{11}$, and so

$$n \equiv \sum_{k=0}^{h} a_k 10^k \equiv \sum_{k=0}^{h} a_k \cdot (-1)^k \equiv s'(n) \pmod{11},$$

from which the conclusion follows.

For (d), the conclusion follows by the facts $1001 = 7 \cdot 11 \cdot 13$ and

$$n = \overline{a_h a_{h-1} \dots a_3} \cdot 1000 + \overline{a_2 a_1 a_0} = \overline{a_h a_{h-1} \dots a_3} \cdot (1001 - 1) + \overline{a_2 a_1 a_0}.$$

For (e), the conclusion follows by the facts $999 = 27 \cdot 37$ and

$$n = \overline{a_h a_{h-1} \dots a_3} \cdot 1000 + \overline{a_2 a_1 a_0} = \overline{a_h a_{h-1} \dots a_3} \cdot (999 + 1) + \overline{a_2 a_1 a_0}.$$

For (f), we note that $10^k \equiv 0 \pmod m$ for $m = 2^k$ or $m = 5^k$. We have

$$n = \overline{a_h \dots a_k} \cdot 10^k + \overline{a_{k-1} \dots a_0},$$

from which the conclusion follows. □

Example 1.50. Perfect squares or not?

(1) Determine all positive integers k such that the k-digit number $11 \dots 1$ is not a perfect square.

(2) Can a 5-digit number consisting only of distinct even digits be a perfect square?

(3) Determine whether $\underbrace{20 \dots 04}_{2004}$ is a perfect square.

Solution: The answers are mostly negative for all these questions.

(1) Clearly, $k = 1$ works. We claim that there are no other answers. Since $\underbrace{11 \dots 1}_{k\ 1\text{'s}} \equiv 11 \equiv 3 \pmod 4$, $\underbrace{11 \dots 1}_{k\ 1\text{'s}}$ is not a perfect square. (Example 1.24 (3)).

(2) The answer is *no*. If n is a 5-digit number consisting only of distinct even digits, then the sum of its digits is equal to $0 + 2 + 4 + 6 + 8 = 20$, which is congruent to 2 modulo 9; hence it is not a perfect square. (Example 1.24 (4).)

(3) The given number is not a perfect square because the sum of its digits is 6, a multiple of 3 but not of 9. (Example 1.24 (4).) □

Example 1.51. [AIME 1984] The integer n is the smallest positive multiple of 15 such that every digit of n is either 0 or 8. Find n.

Solution: An integer n is divisible by 15 if and only if it is divisible by both 3 and 5. By Proposition 1.44 (a) and (f), the answer is $n = 8880$. □

Example 1.52. Determine the number of five-digit positive integers $\overline{abcde}$ ($a, b, c, d,$ and e not necessarily distinct) such that the sum of the three-digit number $\overline{abc}$ and the two-digit number $\overline{de}$ is divisible by 11.

Solution: The answer is 8181.

Note that

$$\overline{abcde} = \overline{abc} \times 100 + \overline{de} = \overline{abc} + \overline{de} + 99 \times \overline{abc}.$$

Hence $\overline{abc} + \overline{de}$ is divisible by 11 if and only if $\overline{abcde}$ is divisible by 11. Note that 99990 is the greatest 5-digit number that is divisible by 11 and that 9999 is the greatest 4-digit number that is divisible by 11. Hence there are $\frac{99990}{11} = 9090$ multiples of 11 that have at most 5 digits, and there are $\frac{9999}{11} = 999$ multiples of 11 that have at most 4 digits. Therefore, there are exactly $9090 - 999 = 8181$ multiples of 11 that have exactly 5 digits. □

Example 1.53. [USAMO 2003] Prove that for every positive integer n there exists an n-digit number divisible by 5^n all of whose digits are odd.

First Solution: We proceed by induction. The property is clearly true for $n = 1$. Assume that $N = a_1a_2\ldots a_n$ is divisible by 5^n and has only odd digits. Consider the numbers

$$\begin{aligned}
N_1 &= 1a_1a_2\ldots a_n = 1 \cdot 10^n + 5^n M = 5^n(1 \cdot 2^n + M),\\
N_2 &= 3a_1a_2\ldots a_n = 3 \cdot 10^n + 5^n M = 5^n(3 \cdot 2^n + M),\\
N_3 &= 5a_1a_2\ldots a_n = 5 \cdot 10^n + 5^n M = 5^n(5 \cdot 2^n + M),\\
N_4 &= 7a_1a_2\ldots a_n = 7 \cdot 10^n + 5^n M = 5^n(7 \cdot 2^n + M),\\
N_5 &= 9a_1a_2\ldots a_n = 9 \cdot 10^n + 5^n M = 5^n(9 \cdot 2^n + M).
\end{aligned}$$

The numbers $1 \cdot 2^n + M, 3 \cdot 2^n + M, 5 \cdot 2^n + M, 7 \cdot 2^n + M, 9 \cdot 2^n + M$ give distinct remainders when divided by 5. Otherwise, the difference of some two of them would be a multiple of 5, which is impossible, because 2^n is not a multiple of 5, nor is the difference of any two of the numbers 1, 3, 5, 7, 9. It follows that one of the numbers N_1, N_2, N_3, N_4, N_5 is divisible by $5^n \cdot 5$, and the induction is complete. □

Second Solution: For an m digit number a, where $m \geq n$, let $\ell(a)$ denote the $m-n$ leftmost digits of a. (That is, we consider $\ell(a)$ as an $(m-n)$-digit number.) It is clear that we can choose a large odd number k such that $a_0 = 5^n \cdot k$ has at least n digits. Assume that a_0 has m_0 digits, where $m_0 \geq n$. Note that a_0 is an odd multiple of 5. Hence the units digit of a_0 is 5.

If the n rightmost digits of a_0 are all odd, then the number $b_0 = a_0 - \ell(a_0) \cdot 10^n$ satisfies the conditions of the problem, because b_0 has only odd digits (the same as the n leftmost digits of a_0) and that b_0 is the difference of two multiples of 5^n.

If there is an even digit among the n rightmost digits of a_0, assume that i_1 is the smallest positive integer such that the i_1th rightmost digit of a_0 is even. Then the number $a_1 = a_0 + 5^n \cdot 10^{i_1-1}$ is a multiple of 5^n with at least n digits. The $(i-1)$th rightmost digit is the same as that of a_0 and the i_1th rightmost digit of a_1 is odd. If the n rightmost digits of a_1 are all odd, then $b_1 = a_1 - \ell(a_1) \cdot 10^n$ satisfies the conditions of the problem. If there is an even digit among the n rightmost digits of a_1, assume that i_2 is the smallest positive integer such that the i_2th rightmost digit of a_1 is even. Then $i_2 > i_1$. Set $a_2 = a_1 + 5^n \cdot 10^{i_2-1}$. We can repeat the above process of checking the rightmost digits of a_2 and eliminate the rightmost even digits of a_2, if there is such a digit among the n rightmost digits of a_2. This process can be repeated at most $n-1$ times because the units digit of a_0 is 5. Thus, we can obtain a number a_k, for some nonnegative integer k, such that a_k is a multiple of 5^n with its n rightmost digits all odd. Then $b_k = a_k - \ell(a_k) \cdot 10^n$ is a number that satisfies the conditions of the problem. □

We can replace the condition of odd digits by any collection of 5 digits that forms a complete set of residue classes modulo 5. In exactly the same way, we can show that for every positive integer n there exists an n-digit number divisible by 2^n all of whose digits form a complete set of residue classes modulo 5.

We close this section with some more discussion on $S(n)$, the sum of the digits of a positive integer n.

Proposition 1.45. Let n be a positive integer, and let $S(n)$ denote the sum of its digits. Then

(a) $9 | S(n) - n$;

(b) $S(n_1 + n_2) \leq S(n_1) + S(n_2)$ (subadditivity property);

(c) $S(n_1 n_2) \leq \min(n_1 S(n_2), n_2 S(n_1))$;

(d) $S(n_1 n_2) \leq S(n_1) S(n_2)$ (submultiplicativity property).

Proof: Part (a) is simply Proposition 1.44 (b). Let us prove (b), (c), and (d). Consider $n_1 = \overline{a_k a_{k-1} \ldots a_0}$, $n_2 = \overline{b_h b_{h-1} \ldots b_0}$, and $n_1 + n_2 = \overline{c_s c_{s-1} \ldots c_0}$.

In order to prove (b), we choose the least t such that $a_i + b_i < 10$ for all $i < t$. Then $a_t + b_t \geq 10$; hence $c_t = a_t + b_t - 10$ and $c_{t+1} \leq a_{t+1} + b_{t+1} + 1$. We obtain

$$\sum_{i=1}^{t+1} c_i \leq \sum_{i=1}^{t+1} a_i + \sum_{i=1}^{t+1} b_i.$$

Continuing this procedure, the conclusion follows.

Because of the symmetry, in order to prove (c) it suffices to prove that $S(n_1 n_2) \leq n_1 S(n_2)$. The last inequality follows by applying the subadditivity property (b) repeatedly. Indeed,

$$S(2n_2) = S(n_2 + n_2) \leq S(n_2) + S(n_2) = 2S(n_2),$$

and after n_1 steps we obtain

$$\begin{aligned} S(n_1 n_2) &= S(\underbrace{n_2 + n_2 + \cdots + n_2}_{n_1 \text{ times}}) \\ &\leq \underbrace{S(n_2) + S(n_2) + \cdots + S(n_2)}_{n_1 \text{ times}} = n_1 S(n_2). \end{aligned}$$

To establish (d), we observe that by (b) and (c),

$$\begin{aligned} S(n_1 n_2) &= S\left(n_1 \sum_{i=0}^{h} b_i 10^i\right) = S\left(\sum_{i=0}^{h} n_1 b_i 10^i\right) \\ &\leq \sum_{i=0}^{h} S(n_1 b_i 10^i) = \sum_{i=0}^{h} S(n_1 b_i) \leq \sum_{i=0}^{h} b_i S(n_1) \\ &= S(n_1) \sum_{i=0}^{h} b_i = S(n_1) S(n_2). \end{aligned}$$

as desired. □

From the proof of Proposition 1.45, we note that it is very important to deal with carryings in working with problems related to the sum of the digits.

Example 1.54. [Russia 1999] In the decimal expansion of n, each digit (except the first digit) is greater than the digit to its left. What is $S(9n)$?

Solution: Write $n = \overline{a_k a_{k-1} \ldots a_0}$. By performing the subtraction

$$\begin{array}{r} a_k\ a_{k-1}\ \ldots\ a_1\ a_0\ 0 \\ -\quad a_k\ \ \ldots\ a_2\ a_1\ a_0 \\ \hline \end{array}$$

we find that the digits of $9n = 10n - n$ are

$$a_k,\ a_{k-1} - a_k,\ \ldots,\ a_1 - a_2,\ a_0 - a_1 - 1,\ 10 - a_0.$$

These digits sum to $10 - 1 = 9$. □

Example 1.55. [Ireland 1996] Find a positive integer n such that $S(n) = 1996S(3n)$.

Solution: Consider

$$n = 1\underbrace{33\ldots3}_{5986\ 3\text{'s}}5.$$

Then

$$3n = 4\underbrace{00\ldots0}_{5986\ 0\text{'s}}5.$$

We have $S(n) = 3 \cdot 5986 + 1 + 5 = 17964 = 1996 \cdot 9 = 1996S(n)$, as desired. □

Example 1.56. Determine whether there is any perfect square that ends in 10 distinct digits.

Solution: The answer is *yes*. We note that

$$\begin{array}{r} 1\,1\,1\,1 \\ \times\ 1\,1\,1\,1 \\ \hline 1\,1\,1\,1 \\ 1\,1\,1\,1\ \ \\ 1\,1\,1\,1\ \ \ \ \\ 1\,1\,1\,1\ \ \ \ \ \ \\ \hline 1\,2\,3\,4\,3\,2\,1 \end{array}$$

Likewise, it is not difficult to see that

$$11111111111^2 = 12345678900987654321$$

is a number that satisfies the conditions of the problem. □

Example 1.57. [IMO 1976] When 4444^{4444} is written in decimal notation, the sum of its digits is A. Let B be the sum of the digits of A. Find the sum of the digits of B.

Solution: The answer is 7.

Let $a = 4444^{4444}$. By our notation, we have $A = S(a)$ and $B = S(A)$, and we want to compute $S(B)$.

First we will show that the sum of the digits of B is fairly small. Note that $4444 < 10000 = 10^4$. Hence

$$a = 4444^{4444} < 10^{4 \cdot 4444} = 10^{17776},$$

and so a cannot have more than 17776 digits. Because each digit is at most a 9, $A = S(a) \le 17776 \cdot 9 = 159984$. Of the natural numbers less than or equal to

159984, the number with the largest digit sum is 99999, and so $B = S(A) \leq 45$. Of the natural numbers less than or equal to 45, the number with the largest digit sum is 39. Hence $S(B) \leq 12$.

By Proposition 1.45 (a), we have

$$S(B) \equiv B \equiv S(A) \equiv A \equiv S(a) \equiv a \equiv 4444^{4444} \pmod 9.$$

It suffices to show that $4444^{4444} \equiv 7 \pmod 9$. Indeed, we have

$$\begin{aligned} 4444^{4444} &\equiv (4+4+4+4)^{4444} \equiv 16^{4444} \equiv (-2)^{4444} \\ &\equiv (-2)^{3\cdot 1481+1} \equiv ((-2)^3)^{1481} \cdot (-2) \equiv (-8)^{1481} \cdot (-2) \\ &\equiv 1 \cdot (-2) \equiv 7 \pmod 9. \end{aligned}$$

□

Floor Function

For a real number x there is a unique integer n such that $n \leq x < n+1$. We say that n is the **greatest integer less than or equal to** x, or the **floor** of x. We write $n = \lfloor x \rfloor$. The difference $x - \lfloor x \rfloor$ is called the **fractional part** of x and is denoted by $\{x\}$. The least integer greater than or equal to x is called the **ceiling** of x and is denoted by $\lceil x \rceil$. If x is an integer, then $\lfloor x \rfloor = \lceil x \rceil$ and $\{x\} = 0$; if x is not an integer, then $\lceil x \rceil = \lfloor x \rfloor + 1$.

We start with four (algebraic) examples to get familiar with these functions.

Example 1.58. [Australia 1999] Solve the following system of equations:

$$\begin{aligned} x + \lfloor y \rfloor + \{z\} &= 200.0, \\ \{x\} + y + \lfloor z \rfloor &= 190.1, \\ \lfloor x \rfloor + \{y\} + z &= 178.8. \end{aligned}$$

Solution: Because $x = \lfloor x \rfloor + \{x\}$ for all real numbers x, adding the three equations gives

$$2x + 2y + 2z = 568.9, \quad \text{or} \quad x + y + z = 284.45.$$

Subtracting each of the three given equations from the last equation gives

$$\begin{aligned} \{y\} + \lfloor z \rfloor &= 84.45, \\ \lfloor x \rfloor + \{z\} &= 94.35, \\ \{x\} + \lfloor y \rfloor &= 105.65. \end{aligned}$$

Therefore $84 = \lfloor 84.45 \rfloor = \lfloor \lfloor z \rfloor + \{y\} \rfloor = \lfloor z \rfloor$, and thus $\lfloor z \rfloor = 84$ and $\{y\} = 0.45$. In the same way we obtain $\lfloor y \rfloor = 105$, and so $y = 105.45$. Similarly, $x = 94.65$ and $z = 84.35$. □

Example 1.59. Determine the distinct numbers in the sequence

$$\left\lfloor \frac{1^2}{2005} \right\rfloor, \left\lfloor \frac{2^2}{2005} \right\rfloor, \ldots, \left\lfloor \frac{2005^2}{2005} \right\rfloor.$$

Solution: For $1 \le i \le 2005$, let

$$a_i = \left\lfloor \frac{i^2}{2005} \right\rfloor.$$

Because $44^2 = 1936 < 2005 < 2025 = 45^2$, $a_1 = a_2 = \cdots = a_{44} = 0$.

For integers m with $m \ge 1002$, since

$$\frac{(m+1)^2}{2005} - \frac{m^2}{2005} = \frac{2m+1}{2005} \ge 1,$$

it follows that $a_m < a_{m+1}$. Hence, $a_{1002}, a_{1003}, \ldots, a_{2005}$ take distinct values.

For positive integers m with $m < 1002$, since

$$\frac{(m+1)^2}{2005} - \frac{m^2}{2005} = \frac{2m+1}{2005} < 1,$$

it follows that $a_{m+1} \le a_m + 1$. Note that this sequence is clearly nondecreasing. We conclude that all the integer values less than a_{1001} have been taken.

Finally, we compute $a_{1001} = 499$ and $a_{1002} = 500$. Therefore, the answer of the problem is $500 + 1004 = 1504$ (namely, values $0, 1, \ldots, 499, a_{1002}, a_{1003}, \ldots, a_{2005}$). □

Example 1.60. [ARML 2003] Find the positive integer n such that $\frac{1}{n}$ is closest to $\{\sqrt{123456789}\}$.

Solution: As shown in the Example 1.56, we have

$$\begin{aligned} 11111.11^2 &= 123456765.4321 < 123456789 \\ &< 123456789.87654321 = 11111.1111^2. \end{aligned}$$

Hence

$$\left\lfloor \sqrt{123456789} \right\rfloor = 11111 \text{ and } \frac{1}{10} < 0.11 < \{\sqrt{123456789}\} < 0.1111 < \frac{1}{9}. \ \square$$

Example 1.61 [AIME 1997] Suppose that a is positive, $\{a^{-1}\} = \{a^2\}$, and $2 < a^2 < 3$. Find the value of $a^{12} - 144a^{-1}$.

Solution: Notice first that the given hypothesis implies that $\{a^{-1}\} = a^{-1}$ (since $1 < a$ and $0 < a^{-1} < 1$) and $\{a^2\} = a^2 - 2$. Hence a must satisfy the equation $a^{-1} = a^2 - 2$, or $a^3 - 2a - 1 = 0$. This factors as

$$(a+1)(a^2 - a - 1) = 0,$$

whose only positive root is $a = \frac{1+\sqrt{5}}{2}$. Now use the relations $a^2 = a + 1$ and $a^3 = 2a + 1$ to calculate

$$a^6 = 8a + 5, \quad a^{12} = 144a + 89, \quad \text{and} \quad a^{13} = 233a + 144,$$

from which it follows that

$$a^{12} - 144a^{-1} = \frac{a^{13} - 144}{a} = 233. \qquad \square$$

Note: By the relation $a^2 = a + 1$, we can show easily that $a^n = F_{n-1}a + F_{n-2}$, where $\{F_n\}_{n=0}^{\infty}$ is the Fibonacci sequence with $F_0 = F_1 = 1$ and $F_{n+1} = F_n + F_{n-1}$ for every positive integer n. This is not surprising if we note that $a^2 = a+1$ is the characteristic equation of the Fibonacci sequence. For more details on this, the reader also can look at chapter five of [4].

Example 1.62. Find all real solutions to the equation

$$4x^2 - 40\lfloor x \rfloor + 51 = 0.$$

Solution: Note that

$$(2x-3)(2x-17) = 4x^2 - 40x + 51 \le 4x^2 - 40\lfloor x \rfloor + 51 = 0,$$

which gives $\frac{3}{2} \le x \le \frac{17}{2}$ and $1 \le \lfloor x \rfloor \le 8$. Then

$$x = \frac{\sqrt{40\lfloor x \rfloor - 51}}{2},$$

so it is necessary to have

$$\lfloor x \rfloor = \left\lfloor \frac{\sqrt{40\lfloor x \rfloor - 51}}{2} \right\rfloor.$$

Testing $\lfloor x \rfloor \in \{1, 2, 3, \ldots, 8\}$ in this equation, we find that $\lfloor x \rfloor$ can equal only 2, 6, 7, or 8. Thus the only solutions for x are $\frac{\sqrt{29}}{2}$, $\frac{\sqrt{189}}{2}$, $\frac{\sqrt{229}}{2}$, and $\frac{\sqrt{269}}{2}$. A quick check confirms that these values work. $\square$

Proposition 1.46. We have the following properties for the floor and the ceiling functions.

(a) If a and b are integers with $b > 0$, and q is the quotient and r is the remainder when a is divided by b, then $q = \lfloor \frac{a}{b} \rfloor$ and $r = \{\frac{a}{b}\} \cdot b$.

(b) For any real number x and any integer n, $\lfloor x + n \rfloor = \lfloor x \rfloor + n$ and $\lceil x + n \rceil = \lceil x \rceil + n$.

(c) If x is an integer, then $\lfloor x \rfloor + \lfloor -x \rfloor = 0$; if x is not an integer, $\lfloor x \rfloor + \lfloor -x \rfloor = 1$.

(d) The floor function is nondecreasing; that is, for $x \le y$, $\lfloor x \rfloor \le \lfloor y \rfloor$.

(e) $\left\lfloor x + \frac{1}{2} \right\rfloor$ rounds x to its nearest integer.

(f) $\lfloor x \rfloor + \lfloor y \rfloor \le \lfloor x + y \rfloor \le \lfloor x \rfloor + \lfloor y \rfloor + 1$.

(g) $\lfloor x \rfloor \cdot \lfloor y \rfloor \le \lfloor xy \rfloor$ for nonnegative real numbers x and y.

(h) For any positive real number x and any positive integer n the number of positive multiples of n not exceeding x is $\lfloor \frac{x}{n} \rfloor$.

(i) For any real number x and any positive integer n,

$$\left\lfloor \frac{\lfloor x \rfloor}{n} \right\rfloor = \left\lfloor \frac{x}{n} \right\rfloor.$$

Proof: The proofs of (a) to (d) are straightforward. We present only the proof of (e) to (i).

For (e) note that if $\{x\} < \frac{1}{2}$, then $\left\lfloor x + \frac{1}{2} \right\rfloor = \lfloor x \rfloor$, which is the integer closest to x; if $\{x\} > \frac{1}{2}$, then $\left\lfloor x + \frac{1}{2} \right\rfloor = \lceil x \rceil$, which is the integer closest to x. This is a very simple but useful trick in computer programming.

For (f), we write $x = \lfloor x \rfloor + \{x\}$ and $y = \lfloor y \rfloor + \{y\}$. The desired result reduces to

$$0 \le \lfloor \{x\} + \{y\} \rfloor \le 1,$$

which is clear since $0 \le \{x\}, \{y\} < 1$.

For (g), we write again $x = \lfloor x \rfloor + \{x\}$ and $y = \lfloor y \rfloor + \{y\}$. Then $\lfloor x \rfloor$, $\lfloor y \rfloor$, $\{x\}$, $\{y\}$ are all nonnegative. It is clear that

$$\begin{aligned} \lfloor xy \rfloor &= \lfloor (\lfloor x \rfloor + \{x\})(\lfloor y \rfloor + \{y\}) \rfloor \\ &= \lfloor \lfloor x \rfloor \lfloor y \rfloor + \lfloor x \rfloor \{y\} + \lfloor y \rfloor \{x\} + \{x\}\{y\} \rfloor \ge \lfloor x \rfloor \lfloor y \rfloor. \end{aligned}$$

For (h), we consider all multiples $1 \cdot n, 2 \cdot n, \ldots, k \cdot n$, where $k \cdot n \le x < (k+1)n$. That is, $k \le \frac{x}{n} < k+1$, and the conclusion follows.

Note that (i) follows immediately from (f), since the multiples of an integer n are integers. □

Furthermore, we extend Proposition 1.46 (f) to the following.

Example 1.63. For real numbers x and y, prove that

$$\lfloor 2x \rfloor + \lfloor 2y \rfloor \ge \lfloor x \rfloor + \lfloor y \rfloor + \lfloor x + y \rfloor.$$

Proof: Write $x = \lfloor x \rfloor + \{x\}$ and $y = \lfloor y \rfloor + \{y\}$. Then

$$\lfloor 2x \rfloor + \lfloor 2y \rfloor = 2\lfloor x \rfloor + \lfloor 2\{x\} \rfloor + 2\lfloor y \rfloor + \lfloor 2\{y\} \rfloor$$

and

$$\lfloor x + y \rfloor = \lfloor x \rfloor + \lfloor y \rfloor + \lfloor \{x\} + \{y\} \rfloor.$$

It suffices to show that

$$\lfloor 2\{x\} \rfloor + \lfloor 2\{y\} \rfloor \ge \lfloor \{x\} + \{y\} \rfloor.$$

By symmetry, we may assume that $\{x\} \ge \{y\}$. Note that $\{x\}$ is nonnegative. We have

$$\lfloor 2\{x\} \rfloor + \lfloor 2\{y\} \rfloor \ge \lfloor 2\{x\} \rfloor \ge \lfloor \{x\} + \{y\} \rfloor,$$

by Proposition 1.46 (e) (since $2\{x\} \ge \{x\} + \{y\}$). □

Proposition 1.46 (e) also has different forms for special values of the variable.

Example 1.64. For a given positive integer n, show that

$$\left\lfloor \sqrt{n} + \frac{1}{2} \right\rfloor = \left\lfloor \sqrt{n - \frac{3}{4}} + \frac{1}{2} \right\rfloor.$$

Proof: Suppose that

$$\left\lfloor \sqrt{n} + \frac{1}{2} \right\rfloor = k \quad \text{and} \quad \left\lfloor \sqrt{n - \frac{3}{4}} + \frac{1}{2} \right\rfloor = m.$$

Then we have $k \le \sqrt{n} + \frac{1}{2} < k + 1$, or $k - \frac{1}{2} \le \sqrt{n} < k + \frac{1}{2}$. Squaring both sides of the last inequality gives

$$k^2 - k + \frac{1}{4} \le n < k^2 + k + \frac{1}{4}.$$

Since n is an integer, we have $k^2 - k + 1 \le n \le k^2 + k$.

Likewise, we have $m \le \sqrt{n - \frac{3}{4}} + \frac{1}{2} < m + 1$, implying that

$$m^2 - m + \frac{1}{4} \le n - \frac{3}{4} < m^2 + m + \frac{1}{4}.$$

Because n is an integer, we again have $m^2 - m + 1 \le n \le m^2 + m$.

Combining the above, we conclude that $m = k$, as desired. □

The graphs of the functions $y = \lfloor x \rfloor$ and $y = \lceil x \rceil$ are typical step functions. Their unique properties allow us to describe some very special sequences.

Example 1.65. [AIME 1985] How many of the first 1000 positive integers can be expressed in the form

$$\lfloor 2x \rfloor + \lfloor 4x \rfloor + \lfloor 6x \rfloor + \lfloor 8x \rfloor,$$

where x is a real number?

Solution: Define the function

$$f(x) = \lfloor 2x \rfloor + \lfloor 4x \rfloor + \lfloor 6x \rfloor + \lfloor 8x \rfloor,$$

and observe that if n is a positive integer, then $f(x + n) = f(x) + 20n$. In particular, this means that if an integer k can be expressed in the form $f(x_0)$ for some real number x_0, then for $n = 1, 2, 3, \ldots$ one can express $k + 20n$ similarly; that is, $k + 20n = f(x_0) + 20n = f(x_0 + n)$. In view of this, one may restrict attention to determining which of the first 20 positive integers are generated by $f(x)$ as x ranges through the half-open interval $(0, 1]$.

Next observe that as x increases, the value of $f(x)$ changes only when either $2x, 4x, 6x$, or $8x$ attains an integral value, and that the change in $f(x)$ is always to a new, higher value. In the interval $(0, 1]$ such changes occur precisely when x is of the form m/n, where $l \le m \le n$ and $n = 2, 4, 6$, or 8. There are 12 such fractions; in increasing order they are

$$\frac{1}{8}, \frac{1}{6}, \frac{1}{4}, \frac{1}{3}, \frac{3}{8}, \frac{1}{2}, \frac{5}{8}, \frac{2}{3}, \frac{3}{4}, \frac{5}{6}, \frac{7}{8}, \text{ and } 1.$$

Therefore, only 12 of the first 20 positive integers can be represented in the desired form. Since $1000 = 50 \cdot 20$, there are $50 \cdot 12 = 600$ positive integers of the desired form. □

Example 1.66. [Gauss] Let p and q be relatively prime integers. Prove that

$$\left\lfloor \frac{p}{q} \right\rfloor + \left\lfloor \frac{2p}{q} \right\rfloor + \cdots + \left\lfloor \frac{(q-1)p}{q} \right\rfloor = \frac{(p-1)(q-1)}{2}.$$

Solution: Since $\gcd(p,q)=1$, $\frac{ip}{q}$ is not an integer. By Proposition 1.46 (c), it follows that

$$\left\lfloor \frac{ip}{q} \right\rfloor + \left\lfloor \frac{(q-i)p}{q} \right\rfloor = p + \left\lfloor \frac{ip}{q} \right\rfloor + \left\lfloor \frac{-ip}{q} \right\rfloor = p-1$$

for $1 \le i \le q-1$. Therefore,

$$\begin{aligned}
&2\left(\left\lfloor \frac{p}{q} \right\rfloor + \left\lfloor \frac{2p}{q} \right\rfloor + \cdots + \left\lfloor \frac{(q-1)p}{q} \right\rfloor\right) \\
&\quad = \left(\left\lfloor \frac{p}{q} \right\rfloor + \left\lfloor \frac{(q-1)p}{q} \right\rfloor\right) + \cdots + \left(\left\lfloor \frac{(q-1)p}{q} \right\rfloor + \left\lfloor \frac{p}{q} \right\rfloor\right) \\
&\quad = (p-1)(q-1),
\end{aligned}$$

from which the desired result follows. □

We can also interpret the above result as the number of lattice points lying inside the triangle bounded by the lines $y=0$, $x=p$, and $y=\frac{qx}{p}$. (A point in the coordinate plane is a lattice point if it has integer coordinates.)

Example 1.67. The sequence

$$\{a_n\}_{n=1}^{\infty} = \{2, 3, 5, 6, 7, 8, 10, \dots\}$$

consists of all the positive integers that are not perfect squares. Prove that

$$a_n = n + \left\lfloor \sqrt{n} + \frac{1}{2} \right\rfloor.$$

First Proof: We claim that

$$\left\lfloor \sqrt{n} + \frac{1}{2} \right\rfloor^2 < n + \left\lfloor \sqrt{n} + \frac{1}{2} \right\rfloor < \left(\left\lfloor \sqrt{n} + \frac{1}{2} \right\rfloor + 1\right)^2. \qquad (\dagger)$$

With our claim, it is clear that among the integers

$$1,\ 2,\ \dots,\ n + \left\lfloor \sqrt{n} + \frac{1}{2} \right\rfloor,$$

there are exactly $\left\lfloor \sqrt{n} + \frac{1}{2} \right\rfloor$ perfect squares, namely, $1^2, 2^2, \dots, \left\lfloor \sqrt{n} + \frac{1}{2} \right\rfloor^2$. Hence

$$n + \left\lfloor \sqrt{n} + \frac{1}{2} \right\rfloor$$

is the nth number in the sequence after all perfect squares have been deleted; that is,

$$a_n = n + \left\lfloor \sqrt{n} + \frac{1}{2} \right\rfloor.$$

Now we prove our claim. Note that $\sqrt{n}$ is either an integer or an irrational number. Hence $\{\sqrt{n}\} \neq \frac{1}{2}$. We consider two cases.

In the first case, we assume that $\{\sqrt{n}\} < \frac{1}{2}$. Set $k = \lfloor \sqrt{n} \rfloor$. Then $k^2 \leq n < \left(k + \frac{1}{2}\right)^2$, or $k^2 < n < k^2 + k + \frac{1}{4}$. Then

$$\left\lfloor \sqrt{n} + \frac{1}{2} \right\rfloor = \lfloor \sqrt{n} \rfloor = k,$$

and the inequality (†) becomes

$$k^2 < n + k < (k+1)^2 = k^2 + 2k + 1,$$

which is evident.

In the second case, we assume that $\{\sqrt{n}\} > \frac{1}{2}$. Again, set $k = \lfloor \sqrt{n} \rfloor$. Then $\left(k + \frac{1}{2}\right)^2 < n < (k+1)^2$, or $k^2 + k + \frac{1}{4} < n < k^2 + 2k + 1$. Then

$$\left\lfloor \sqrt{n} + \frac{1}{2} \right\rfloor = \lfloor \sqrt{n} \rfloor + 1 = k + 1,$$

and the inequality (†) becomes

$$(k+1)^2 < n + k + 1 < (k+2)^2 = k^2 + 4k + 4,$$

which is also evident.

Combining the last two cases, we have shown that our claim is always true, and our proof is complete. □

The second proof reveals the origin of this closed form of a_n.

Second Proof: Consider the sequence

$$\{b_n\}_{n=1}^{\infty} = \{1, 1; 2, 2, 2, 2; 3, 3, 3, 3, 3, 3; \ldots\}.$$

We note that

$$a_n - b_n = n$$

for all positive integers n. It is clear that there are are exactly $(n+1)^2 - n^2 - 1 = 2n$ non-perfect squares strictly between two consecutive perfect squares n^2 and $(n+1)^2$. It suffices to show that

$$b_n = \left\lfloor \sqrt{n} + \frac{1}{2} \right\rfloor.$$

If $b_n = k$, it is in the kth group and is preceded by at least $k-1$ groups containing $2+4+\cdots+2(k-1)$ terms. Considering also the fact that there are $n-1$ terms before b_n, we conclude that

$$2+4+\cdots+2(b_n-1) \le n-1.$$

Moreover, b_n is the largest integer satisfying this inequality. Thus b_n is the largest integer satisfying the inequality $b_n(b_n-1) \le n-1$; that is,

$$b_n = \left\lfloor \frac{1+\sqrt{4n-3}}{2} \right\rfloor = \left\lfloor \sqrt{n - \frac{3}{4}} + \frac{1}{2} \right\rfloor = \left\lfloor \sqrt{n} + \frac{1}{2} \right\rfloor,$$

by Example 1.64. □

Theorem 1.47. [Beatty's Theorem] Let α and β be two positive irrational real numbers such that

$$\frac{1}{\alpha} + \frac{1}{\beta} = 1.$$

The sets

$$\{a_n\}_{n=1}^{\infty} = \{\lfloor \alpha \rfloor, \lfloor 2\alpha \rfloor, \lfloor 3\alpha \rfloor, \dots\} \quad \text{and} \quad \{b_n\}_{n=1}^{\infty} = \{\lfloor \beta \rfloor, \lfloor 2\beta \rfloor, \lfloor 3\beta \rfloor, \dots\}$$

form a partition of the set of positive integers; that is, $\{a_n\}_{n=1}^{\infty}$ and $\{b_n\}_{n=1}^{\infty}$ are nonintersecting sets with their union equal to the set of all positive integers.

Proof: We first show that they are nonintersecting. We proceed indirectly by assuming the contrary, that is, we assume that there are indices i and j such that $k = a_i = b_j = \lfloor i\alpha \rfloor = \lfloor j\beta \rfloor$. Since both $i\alpha$ and $j\beta$ are irrational, it follows that

$$k < i\alpha < k+1, \; k < j\beta < k+1,$$

or

$$\frac{i}{k+1} < \frac{1}{\alpha} < \frac{i}{k} \quad \text{and} \quad \frac{j}{k+1} < \frac{1}{\beta} < \frac{j}{k}.$$

Adding these two inequalities gives

$$\frac{i+j}{k+1} < \frac{1}{\alpha} + \frac{1}{\beta} = 1 < \frac{i+j}{k},$$

or $k < i + j < k + 1$, which is impossible. Hence our assumption was wrong and these two sequences do not intersect.

We next prove that every positive integer appears in one of the two sequences. We again approach indirectly by assuming that there is a positive integer k that does not appear in these two sequences. It follows that there are indices i and j such that

$$i\alpha < k, \quad (i+1)\alpha > k+1, \quad j\beta < k, \quad (j+1)\beta > k+1,$$

or

$$\frac{i}{k} < \frac{1}{\alpha} < \frac{i+1}{k+1} \quad \text{and} \quad \frac{j}{k} < \frac{1}{\beta} < \frac{j+1}{k+1}.$$

Adding the last two inequalities gives

$$\frac{i+j}{k} < \frac{1}{\alpha} + \frac{1}{\beta} = 1 < \frac{i+j+2}{k+1},$$

implying that $i + j < k$ and $k + 1 < i + j + 2$, and so $i + j < k < i + j + 1$, which is again impossible. Hence our assumption was wrong and every positive integer appears in exactly one of the two sequences. □

Example 1.68a. [USAMO 1981] For a positive number x, prove that

$$\lfloor x \rfloor + \frac{\lfloor 2x \rfloor}{2} + \frac{\lfloor 3x \rfloor}{3} + \cdots + \frac{\lfloor nx \rfloor}{n} \le \lfloor nx \rfloor.$$

Indeed, we have a more general result. By Proposition 1.46 (f), Example 1.68a is a special case of Example 1.68b by setting $a_i = -\lfloor ix \rfloor$.

Example 1.68b. [APMO 1999] Let $a_1, a_2, \ldots$ be a sequence of real numbers satisfying

$$a_{i+j} \le a_i + a_j$$

for all $i, j = 1, 2, \ldots$. Prove that

$$a_1 + \frac{a_2}{2} + \frac{a_3}{3} + \cdots + \frac{a_n}{n} \ge a_n$$

for all positive integers n.

First Proof: We use strong induction. The base cases for $n = 1$ and 2 are trivial. Now assume that the statement is true for $n \le k$ for some positive integer $k \ge 2$.

That is,

$$\begin{aligned} a_1 &\ge a_1, \\ a_1 + \frac{a_2}{2} &\ge a_2, \\ &\vdots \\ a_1 + \frac{a_2}{2} + \cdots + \frac{a_k}{k} &\ge a_k. \end{aligned}$$

Adding all the inequalities gives

$$ka_1 + (k-1)\frac{a_2}{2} + \cdots + \frac{a_k}{k} \ge a_1 + a_2 + \cdots + a_k.$$

Adding $(a_1 + a_2 + \cdots + a_k)$ to both sides of the last inequality yields

$$\begin{aligned} (k+1)\left(a_1 + \frac{a_2}{2} + \cdots + \frac{a_k}{k}\right) &\ge (a_1 + a_k) + (a_2 + a_{k-1}) + \cdots + (a_k + a_1) \\ &\ge ka_{k+1}. \end{aligned}$$

Dividing both sides of the last inequality by $(k+1)$ gives

$$a_1 + \frac{a_2}{2} + \cdots + \frac{a_k}{k} \ge \frac{ka_{k+1}}{k+1},$$

or

$$a_1 + \frac{a_2}{2} + \cdots + \frac{a_k}{k} + \frac{a_{k+1}}{k+1} \ge a_{k+1}.$$

This completes the induction and we are done. □

Second Proof: [By Andreas Kaseorg] We can extend the condition by induction to $a_{i_1+i_2+\cdots+i_k} \le a_{i_1} + a_{i_2} + \cdots + a_{i_k}$. We apply a combinatorial argument.

A **permutation** is a change in position within a collection. More precisely, if S is a set, then a permutation of S is a one-to-one function π that maps S onto itself. If $S = \{x_1, x_2, \ldots, x_n\}$ is a finite set, then we may denote a permutation π of S by $(y_1, y_2, \ldots, y_n)$, where $y_k = \pi(x_k)$. An ordered k-tuple $(x_{i_1}, x_{i_2}, \ldots, x_{i_k})$ is a **k-cycle** of π if $\pi(x_{i_1}) = x_{i_2}$, $\pi(x_{i_2}) = x_{i_3}$, $\ldots$, and $\pi(x_{i_k}) = x_1$. Let S_n denote the set of permutations of n elements. For an element π in S_n, define $f(\pi, k)$ to be the number of k-cycles in π. Clearly,

$$1 \cdot f(\pi, 1) + 2 \cdot f(\pi, 2) + \cdots + n \cdot f(\pi, n) = n,$$

since both sides count the number of elements in the permutation π. Note also that $\sum_{\pi \in S_n} f(\pi, k)$ is the total number of k-cycles in all permutations on n elements, which is $\binom{n}{k}(k-1)!(n-k)! = \frac{n!}{k}$; that is,

$$\sum_{\pi \in S_n} f(\pi, k) = \binom{n}{k}(k-1)!(n-k)! = \frac{n!}{k}. \qquad (*)$$

This is because

(a) we have $\binom{n}{k}$ ways to choose k elements as the elements of a k-cycle;

(b) we have $(k-1)!$ ways to form a k-cycle using the k chosen elements;

(c) we have $(n-k)!$ ways to permute the $n-k$ unchosen elements to complete the permutation of all n elements.

Therefore, by $(*)$, we have

$$\begin{aligned} a_1 + \frac{a_2}{2} + \frac{a_3}{3} + \cdots + \frac{a_n}{n} & \\ &= \frac{1}{n!} \sum_{\pi \in S_n} [f(\pi, 1)a_1 + f(\pi, 2)a_2 + \cdots + f(\pi, n)a_n] \\ &\geq \frac{1}{n!} \sum_{\pi \in S_n} a_{1 \cdot f(\pi,1) + 2 \cdot f(\pi,2) + \cdots + n \cdot f(\pi,n)} \\ &= \frac{1}{n!} \sum_{\pi \in S_n} a_n = a_n, \end{aligned}$$

because there are exactly $n!$ elements in S_n (since there are $n!$ permutations of n elements). □

As shown in Examples 1.68a and 1.68b, many interesting and challenging problems related to the floor and ceiling functions have close ties to their special functional properties. We leave most of them to the sequel of this book: *105 Diophantine Equations and Integer Function Problems*. We close this section by introducing the well-known **Hermite identity**.

Proposition 1.48. [Hermite Identity] Let x be a real number, and let n be a positive integer. Then

$$\lfloor x \rfloor + \left\lfloor x + \frac{1}{n} \right\rfloor + \left\lfloor x + \frac{2}{n} \right\rfloor + \cdots + \left\lfloor x + \frac{n-1}{n} \right\rfloor = \lfloor nx \rfloor.$$

Proof: If x is an integer, then the result is clearly true. We assume that x is not an integer; that is, $0 < \{x\} < 1$. Then there exists $1 \leq i \leq n-1$ such that

$$\{x\} + \frac{i-1}{n} < 1 \quad \text{and} \quad \{x\} + \frac{i}{n} \geq 1, \qquad (*)$$

that is,

$$\frac{n-i}{n} \le \{x\} < \frac{n-i+1}{n}. \qquad (**)$$

By (*), we have

$$\lfloor x \rfloor = \left\lfloor x + \frac{1}{n} \right\rfloor = \cdots = \left\lfloor x + \frac{i-1}{n} \right\rfloor$$

and

$$\left\lfloor x + \frac{i}{n} \right\rfloor = \cdots = \left\lfloor x + \frac{n-1}{n} \right\rfloor = \lfloor x \rfloor + 1,$$

and so

$$\lfloor x \rfloor + \left\lfloor x + \frac{1}{n} \right\rfloor + \left\lfloor x + \frac{2}{n} \right\rfloor + \cdots + \left\lfloor x + \frac{n-1}{n} \right\rfloor$$
$$= i\lfloor x \rfloor + (n-i)(\lfloor x \rfloor + 1) = n\lfloor x \rfloor + n - i.$$

On the other hand, by (**), we obtain

$$n\lfloor x \rfloor + n - i \le n\lfloor x \rfloor + n\{x\} = nx < n\lfloor x \rfloor + n - i + 1,$$

implying that $\lfloor nx \rfloor = n\lfloor x \rfloor + n - i$.

Combining the above observations, we have

$$\lfloor x \rfloor + \left\lfloor x + \frac{1}{n} \right\rfloor + \left\lfloor x + \frac{2}{n} \right\rfloor + \cdots + \left\lfloor x + \frac{n-1}{n} \right\rfloor = n\lfloor x \rfloor + n - i$$
$$= \lfloor nx \rfloor.$$

□

Example 1.69. [AIME 1991] Suppose that r is a real number for which

$$\left\lfloor r + \frac{19}{100} \right\rfloor + \left\lfloor r + \frac{20}{100} \right\rfloor + \cdots + \left\lfloor r + \frac{91}{100} \right\rfloor = 546.$$

Find $\lfloor 100r \rfloor$.

Solution: The given sum has $91 - 19 + 1 = 73$ terms, each of which equals either $\lfloor r \rfloor$ or $\lfloor r \rfloor + 1$. But $73 \cdot 7 < 546 < 73 \cdot 8$, and so it follows that $\lfloor x \rfloor = 7$. Because $546 = 73 \cdot 7 + 35$, the first 38 terms take the value 7 and the last 35 terms take the value 8; that is,

$$\left\lfloor r + \frac{56}{100} \right\rfloor = 7 \quad \text{and} \quad \left\lfloor r + \frac{57}{100} \right\rfloor = 8.$$

It follows that $7.43 \le r < 7.44$ and hence that $\lfloor 100r \rfloor = 743$. □

Example 1.70. [IMO 1968] Let x be a real number. Prove that

$$\sum_{k=0}^{\infty}\left\lfloor\frac{x+2^k}{2^{k+1}}\right\rfloor = \lfloor x\rfloor.$$

Solution: Setting $n = 2$ in Hermite's identity gives

$$\lfloor x\rfloor + \left\lfloor x+\frac{1}{2}\right\rfloor = \lfloor 2x\rfloor,$$

or

$$\left\lfloor x+\frac{1}{2}\right\rfloor = \lfloor 2x\rfloor - \lfloor x\rfloor.$$

Repeatedly applying the last identity gives

$$\sum_{k=0}^{\infty}\left\lfloor\frac{x+2^k}{2^{k+1}}\right\rfloor = \sum_{k=0}^{\infty}\left\lfloor\frac{x}{2^{k+1}}+\frac{1}{2}\right\rfloor = \sum_{k=0}^{\infty}\left(\left\lfloor\frac{x}{2^k}\right\rfloor - \left\lfloor\frac{x}{2^{k+1}}\right\rfloor\right) = \lfloor x\rfloor,$$

as desired. □

Legendre's Function

We use Proposition 1.46 (h) to develop some interesting results.

Let p be a prime. For any positive integer n, let $e_p(n)$ be the exponent of p in the prime factorization of $n!$. The arithmetic function e_p is called the **Legendre function** associated with the prime p.

The following result gives a formula for the computation of $e_p(n)$.

Proposition 1.49. [Legendre's Formula] For any prime p and any positive integer n,

$$e_p(n) = \sum_{i\geq 1}\left\lfloor\frac{n}{p^i}\right\rfloor = \left\lfloor\frac{n}{p}\right\rfloor + \left\lfloor\frac{n}{p^2}\right\rfloor + \left\lfloor\frac{n}{p^3}\right\rfloor + \cdots.$$

We note that this sum is a finite one, because for large m, $n < p^{m+1}$ and $\left\lfloor\frac{n}{p^{m+1}}\right\rfloor = 0$. Let m be the least positive integer such that $n < p^{m+1}$; that is, $m = \left\lfloor\frac{\ln n}{\ln p}\right\rfloor$. It suffices to show that

$$e_p(n) = \left\lfloor\frac{n}{p}\right\rfloor + \left\lfloor\frac{n}{p^2}\right\rfloor + \cdots + \left\lfloor\frac{n}{p^m}\right\rfloor.$$

We present two closely related proofs. The first is written in the language of number theory, the second in the language of combinatorics.

First Proof: For $n < p$ it is clear that $e_p(n) = 0$. If $n \geq p$, then in order to determine $e_p(n)$ we need to consider only the multiples of p in the product $n! = 1 \cdot 2 \cdots n$; that is, $(1 \cdot p)(2 \cdot p) \cdots (kp) = p^k k!$, where $k = \left\lfloor \frac{n}{p} \right\rfloor$ by Proposition 1.46 (h). Hence

$$e_p(n) = \left\lfloor \frac{n}{p} \right\rfloor + e_p\left(\left\lfloor \frac{n}{p} \right\rfloor\right).$$

Replacing n by $\left\lfloor \frac{n}{p} \right\rfloor$ and taking into account Proposition 1.46 (i), we obtain

$$e_p\left(\left\lfloor \frac{n}{p} \right\rfloor\right) = \left\lfloor \frac{\left\lfloor \frac{n}{p} \right\rfloor}{p} \right\rfloor + e_p\left(\left\lfloor \frac{\left\lfloor \frac{n}{p} \right\rfloor}{p} \right\rfloor\right) = \left\lfloor \frac{n}{p^2} \right\rfloor + e_p\left(\left\lfloor \frac{n}{p^2} \right\rfloor\right).$$

Continuing this procedure we get

$$e_p\left(\left\lfloor \frac{n}{p^2} \right\rfloor\right) = \left\lfloor \frac{n}{p^3} \right\rfloor + e_p\left(\left\lfloor \frac{n}{p^3} \right\rfloor\right),$$

$$\vdots$$

$$e_p\left(\left\lfloor \frac{n}{p^{m-1}} \right\rfloor\right) = \left\lfloor \frac{n}{p^m} \right\rfloor + e_p\left(\left\lfloor \frac{n}{p^m} \right\rfloor\right) = \left\lfloor \frac{n}{p^m} \right\rfloor.$$

Summing up the relations above yields the desired result. □

Second Proof: For each positive integer i, define t_i such that $p^{t_i} \| i$. Because p is prime, we have $p^{t_1+t_2+\cdots+t_n} \| n!$, or $t = t_{n!} = t_1 + t_2 + \cdots + t_n$. On the other hand, $\left\lfloor \frac{n}{p^k} \right\rfloor$ counts all multiples of p^k that are less than or equal to n exactly once. Thus the number $i = p^{t_i} \cdot a$ (with a and p relatively prime) is counted t_i times in the sum

$$\left\lfloor \frac{n}{p} \right\rfloor + \left\lfloor \frac{n}{p^2} \right\rfloor + \cdots + \left\lfloor \frac{n}{p^m} \right\rfloor,$$

namely, in the terms $\left\lfloor \frac{n}{p} \right\rfloor, \left\lfloor \frac{n}{p^2} \right\rfloor, \ldots, \left\lfloor \frac{n}{p^i} \right\rfloor$. Therefore, for each $1 \leq i \leq n$, the number i contributes t_i in both

$$\left\lfloor \frac{n}{p} \right\rfloor + \left\lfloor \frac{n}{p^2} \right\rfloor + \cdots + \left\lfloor \frac{n}{p^m} \right\rfloor$$

and

$$t_1 + t_2 + \cdots + t_n.$$

Hence

$$t = t_1 + t_2 + \cdots + t_n = \left\lfloor \frac{n}{p} \right\rfloor + \left\lfloor \frac{n}{p^2} \right\rfloor + \cdots + \left\lfloor \frac{n}{p^m} \right\rfloor.$$

In a more formal language, consider the matrix $\mathbf{M} = (x_{i,j})$ with m rows and n columns, where m is the smallest integer such that $p^m > n$. We define

$$x_{i,j} = \begin{cases} 1 \text{ if } p^i \text{ divides } j, \\ 0 \text{ otherwise.} \end{cases}$$

Then the number of 1's in the jth column of the matrix $\mathbf{M}$ is t_j, implying that the column sums of $\mathbf{M}$ are $t_1, t_2, \ldots, t_n$. Hence the sum of all of the entries in $\mathbf{M}$ is t. On the other hand, the 1's in the ith row denote the numbers that are multiples of p^i. Consequently, the sum of the entries in the ith row is $\left\lfloor \frac{n}{p^i} \right\rfloor$. Thus the sum of the entries in $\mathbf{M}$ is also $\sum_{i=1}^{m} \left\lfloor \frac{n}{p^i} \right\rfloor$. It follows that

$$t = \left\lfloor \frac{n}{p} \right\rfloor + \left\lfloor \frac{n}{p^2} \right\rfloor + \cdots + \left\lfloor \frac{n}{p^m} \right\rfloor,$$

as desired. □

Example 1.71. Let s and t be positive integers such that

$$7^s \| 400! \quad \text{and} \quad 3^t \| ((3!)!)!.$$

Compute $s + t$.

Solution: The answer is 422.

Note that $((3!)!)! = (6!)! = 720!$. Applying Legendre's formula, we have

$$s = e_7(400) = \left\lfloor \frac{400}{7} \right\rfloor + \left\lfloor \frac{400}{7^2} \right\rfloor + \left\lfloor \frac{400}{7^3} \right\rfloor = 57 + 8 + 1 = 66$$

and

$$\begin{aligned} t = e_3(720) &= \left\lfloor \frac{720}{3} \right\rfloor + \left\lfloor \frac{720}{3^2} \right\rfloor + \left\lfloor \frac{720}{3^3} \right\rfloor + \left\lfloor \frac{720}{3^4} \right\rfloor + \left\lfloor \frac{720}{3^5} \right\rfloor \\ &= 240 + 80 + 26 + 8 + 2 = 356, \end{aligned}$$

and so $s + t = 356 + 66 = 422$. □

Example 1.72. The decimal representation of 2005! ends in m zeros. Find m.

Solution: It is equivalent to compute m such that $10^m \| 2005!$. Since $10^m = 2^m 5^m$, we have $m = \min\{e_2(2005!), e_5(2005!)\}$. Because $2 < 5$, we have

$$m = e_5(2005!) = \left\lfloor \frac{2005}{5} \right\rfloor + \left\lfloor \frac{2005}{25} \right\rfloor + \left\lfloor \frac{2005}{125} \right\rfloor + \left\lfloor \frac{2005}{625} \right\rfloor = 500.$$

The answer is 500. □

Example 1.73. [HMMT 2003] Find the smallest n such that $n!$ ends in 290 zeros.

Solution: As shown in the solution of Example 1.72, we need to find the smallest n such that

$$290 = e_5(n) = \left\lfloor \frac{n}{5} \right\rfloor + \left\lfloor \frac{n}{5^2} \right\rfloor + \left\lfloor \frac{n}{5^3} \right\rfloor + \cdots,$$

which is roughly a geometric series (by taking away the floor function) whose sum is represented approximately by $\frac{n/5}{1-1/5}$. Solving

$$290 \approx \frac{\frac{n}{5}}{1 - \frac{1}{5}},$$

we estimate $n = 1160$, and this gives us $e_5(1160) = 288$. Adding 10 to the value of $n = 1160$ gives the necessary two additional factors of 5 (from 1165 and 1170), and so the answer is 1170. □

Example 1.74. Let m and n be positive integers. Prove that

(1) $m! \cdot (n!)^m$ divides $(mn)!$.

(2) $m!n!(m+n)!$ divides $(2m)!(2n)!$.

Proof: We present a common technique in this proof.

(1) Let p be a prime. Let x and y be nonnegative integers such that $p^x \| m! \cdot (n!)^m$ and $p^y \| (mn)!$. It suffices to show that $x \le y$. Note that $x = e_p(m) + me_p(n)$ and $y = e_p(mn)$. It suffices to show that

$$\sum_{i=1}^{\infty} \left\lfloor \frac{mn}{p^i} \right\rfloor \ge \sum_{i=1}^{\infty} \left\lfloor \frac{m}{p^i} \right\rfloor + m \sum_{i=1}^{\infty} \left\lfloor \frac{n}{p^i} \right\rfloor.$$

If $p > n$, then the second summand on the right-hand side is 0 and the inequality is clearly true. We assume that $p \le n$. Let s be the positive integer such that $p^s \le n < p^{s+1}$. By Proposition 1.46 (g), we have

$$\sum_{i=1}^{\infty}\left\lfloor \frac{mn}{p^i}\right\rfloor = \sum_{i=1}^{s}\left\lfloor m\cdot\frac{n}{p^i}\right\rfloor + \sum_{i=1}^{\infty}\left\lfloor \frac{m}{p^i}\cdot\frac{n}{p^s}\right\rfloor$$
$$\ge m\sum_{i=1}^{s}\left\lfloor \frac{n}{p^i}\right\rfloor + \sum_{i=1}^{\infty}\left\lfloor \frac{m}{p^i}\right\rfloor\left\lfloor \frac{n}{p^s}\right\rfloor$$
$$\ge m\sum_{i=1}^{\infty}\left\lfloor \frac{n}{p^i}\right\rfloor + \sum_{i=1}^{\infty}\left\lfloor \frac{m}{p^i}\right\rfloor,$$

as desired.

(2) The proof is very similar to that of (1). We leave it to the reader. □

Note: It is also to find combinatorial proofs for these facts. For example, there are

$$\frac{(mn)!}{m!(n!)^m}$$

ways to split mn people into m groups of n, implying (1).

Example 1.75. Let k and n be positive integers. Prove that

$$(k!)^{k^n+k^{n-1}+\cdots+k+1} \mid (k^{n+1})!.$$

Proof: For every i with $0 \le i \le n$, setting $(n, m) = (k, k^i)$ in Example 1.74 (1) gives

$$k! \mid k!,\ k!(k!)^k \mid (k^2)!,\ (k^2)!(k!)^{k^2} \mid (k^3)!,\ \ldots,\ (k^n)!(k!)^{k^n} \mid (k^{n+1})!.$$

Multiplying this together gives

$$k!k!(k^2!)(k^3!)\cdots(k^n)!k!^{k+k^2+\cdots+k^n} \mid k!(k^2!)(k^3!)\cdots(k^{n+1})!,$$

from which the desired result follows. □

Example 1.76. Let $n > 2$ be a composite number. Prove that not all of the terms in the sequence

$$\binom{n}{1},\binom{n}{2},\ldots,\binom{n}{n-1}$$

are divisible by n.

Proof: Let p be a prime divisor of n, and let s be the integer such that $p^s \le n < p^{s+1}$. We show that

$$n \nmid \binom{n}{p^s} = \frac{n!}{(p^s)!(n-p^s)!}.$$

Since $p \mid n$, it suffices to show that $p \nmid \binom{n}{p^s}$. Suppose that $p^k \| \binom{n}{p^s}$. Then

$$k = e_p(n) - e_p(p^s) - e_p(n-p^s).$$

It suffices to show that $k = 0$. By Legendre's formula, we have

$$\begin{aligned} k &= \sum_{i\ge1}\left\lfloor\frac{n}{p^i}\right\rfloor - \sum_{i\ge1}\left\lfloor\frac{p^s}{p^i}\right\rfloor - \sum_{i\ge1}\left\lfloor\frac{n-p^s}{p^i}\right\rfloor \\ &= \sum_{i=1}^{s}\left\lfloor\frac{n}{p^i}\right\rfloor - \sum_{i=1}^{s}\left\lfloor\frac{p^s}{p^i}\right\rfloor - \sum_{i=1}^{s}\left\lfloor\frac{n-p^s}{p^i}\right\rfloor \\ &= \sum_{i=1}^{s}\left\lfloor\frac{n}{p^i}\right\rfloor - \sum_{i=1}^{s}\left\lfloor\frac{p^s}{p^i}\right\rfloor - \sum_{i=1}^{s}\left\lfloor\frac{n}{p^i}\right\rfloor + \sum_{i=1}^{s}\left\lfloor\frac{p^s}{p^i}\right\rfloor = 0, \end{aligned}$$

since $\left\lfloor\frac{p^s}{p^i}\right\rfloor$ are integers for each $1 \le i \le s$. □

Legendre's formula is a great tool in combinatorial number theory. It helps to establish two important theorems of Lucas and Kummer. We will discuss them in detail in the sequel to this book – *107 Combinatorial Number Theory Problems*. The reader can also look at chapter three of [4].

Fermat Numbers

Trying to find all primes of the form $2^m + 1$, Fermat noticed that m must be a power of 2. Indeed, if m were equal to $k \cdot h$ with k an odd integer greater than 1, then

$$2^m + 1 = (2^h)^k + 1 = (2^h + 1)(2^{h(k-1)} - 2^{h(k-2)} + \cdots - 2^h + 1),$$

and so $2^m + 1$ would not be a prime.

The integers $f_n = 2^{2^n} + 1$, $n \ge 0$, are called **Fermat numbers**. We have

$$f_0 = 3,\ f_1 = 5,\ f_2 = 17,\ f_3 = 257,\ f_4 = 65537, \text{ and } f_5 = 4294967297.$$

After checking that these five numbers are primes, Fermat conjectured that f_n is a prime for all n. But Euler proved that $641 \mid f_5$. His argument was the following:

$$\begin{aligned} f_5 = 2^{32} + 1 &= 2^{28}(5^4 + 2^4) - (5 \cdot 2^7)^4 + 1 = 2^{28} \cdot 641 - (640^4 - 1) \\ &= 641(2^{28} - 639(640^2 + 1)). \end{aligned}$$

It is still unknown whether there are infinitely many prime Fermat numbers (Fermat primes). The answer to this question is important because Gauss proved that a regular polygon $Q_1Q_2\ldots Q_n$ can be constructed using a straightedge and compass if and only if $n = 2^h p_0 \cdots p_k$, where $k \geq 0$, $p_0 = 1$, and $p_1, \ldots, p_k$ are distinct Fermat primes. Gauss was the first to construct such a polygon for $n = 17$. It is also unknown whether there are infinitely many composite Fermat numbers. (Well, the *good* thing is that the answer to one of these two questions must be positive ☺.)

Example 1.77. For positive integers m and n with $m > n$, f_n divides $f_m - 2$.

Proof: By repeatedly applying the difference of squares formula $a^2 - b^2 = (a-b)(a+b)$, it is not difficult to show that

$$f_m - 2 = f_{m-1}f_{m-2}\cdots f_1 f_0,$$

from which the desired result follows. □

Example 1.78. For distinct positive integers m and n, f_m and f_n are relatively prime.

Proof: By the first example, we have $\gcd(f_m, f_n) = \gcd(f_n, 2) = 1$. □

This result also is a special case of Example 1.22.

Example 1.79. Prove that for all positive integers n, f_n divides $2^{f_n} - 2$.

Proof: We have

$$2^{f_n} - 2 = 2\left(2^{2^{2^n}} - 1\right) = 2\left[\left(2^{2^n}\right)^{2^{2^n - n}} - 1\right].$$

Clearly, $2^{2^n - n}$ is even. Note that for an even positive integer $2m$, $x^{2m} - 1$ is divisible by $x+1$. Hence $x+1$ divides $x^{2^{2^n-n}} - 1$. Setting $x = 2^{2^n}$ leads to the desired conclusion. □

The result in Example 1.79 shows that $2^{f_n} \equiv 2 \pmod{f_n}$, which gives another counterexample to the converse of Fermat's little theorem. That is, $2^{f_5} \equiv 2 \pmod{f_5}$ but f_5 is not a prime.

Mersenne Numbers

The integers $M_n = 2^n - 1$, $n \geq 1$, are called **Mersenne numbers**. It is clear that if n is composite, then so is M_n. Hence M_k is a prime only if k is a prime. Moreover, if $n = ab$, where a and b are integers greater than 1, then M_a and M_b both divide M_n. But there are primes n for which M_n is composite. For example, $47 \mid M_{23}$, $167 \mid M_{83}$, $263 \mid M_{13}$, and so on.

Theorem 1.50. Let p be an odd prime and let q be a prime divisor of M_p. Then $q = 2kp + 1$ for some positive integer k.

Proof: From the congruence $2^p \equiv 1 \pmod{q}$ and from the fact that p is a prime, by Proposition 1.30, it follows that p is the least positive integer satisfying this property. By using Fermat's little theorem, we have $2^{q-1} \equiv 1 \pmod{q}$, hence $p \mid (q-1)$, by Proposition 1.30 again. But $q-1$ is an even integer, so $q-1 = 2kp$ and the conclusion follows. □

Perfect Numbers

An integer $n \geq 2$ is called **perfect** if the sum of its divisors is equal to $2n$; that is, $\sigma(n) = 2n$. For example, the numbers 6, 28, 496 are perfect. The perfect numbers are closely related to Mersenne numbers. We first introduce a famous result on even perfect numbers. The "if" part belongs to Euclid and the "only if" part is due to Euler.

Theorem 1.51. An even positive integer n is perfect if and only if $n = 2^{k-1}M_k$ for some positive integer k for which M_k is a prime.

Proof: First we show the "if" part. Assume that $n = 2^{k-1}(2^k - 1)$, where $M_k = 2^k - 1$ is prime. Because $\gcd(2^{k-1}, 2^k - 1) = 1$ and the fact that σ is a multiplicative function, it follows that

$$\sigma(n) = \sigma(2^{k-1})\sigma(2^k - 1) = (2^k - 1) \cdot 2^k = 2n;$$

that is, n is perfect.

Second we show the "only if" part. Assume that n is an even perfect number. Let $n = 2^t u$, where $t \geq 0$ and u is odd. Because n is perfect, we have $\sigma(n) = 2n$; hence $\sigma(2^t u) = 2^{t+1}u$. Using again that σ is multiplicative, we get

$$2^{t+1}u = \sigma(2^t u) = \sigma(2^t)\sigma(u) = (2^{t+1} - 1)\sigma(u).$$

Because $\gcd(2^{t+1}-1, 2^{t+1}) = 1$, it follows that $2^{t+1} \mid \sigma(u)$; hence $\sigma(u) = 2^{t+1}v$ for some positive integer v. We obtain $u = (2^{t+1} - 1)v$.

The next step is to show that $v = 1$. Assume to the contrary that $v > 1$. Then

$$\sigma(u) \geq 1 + v + 2^{t+1} - 1 + v(2^{t+1} - 1) = (v + 1)2^{t+1} > v \cdot 2^{t+1} = \sigma(u),$$

a contradiction. We get $v = 1$, hence $u = 2^{t+1} - 1 = M_{t+1}$ and $\sigma(u) = 2^{t+1}$. If M_{t+1} is not a prime, then $\sigma(u) > 2^{t+1}$, which is impossible. Finally, $n = 2^{k-1}M_k$, where $k = t + 1$. □

Since M_k is a prime only if k is a prime, we can reword Theorem 1.51 as

An even positive integer n is perfect if and only if $n = 2^{k-1}M_k$ for some prime p for which M_p is a prime.

Theorem 1.51 sets up a one-to-one correspondence between the prime Mersenne numbers and the even perfect numbers. The following are two simple results related to odd perfect numbers.

Theorem 1.52. If n is an odd perfect number, then the prime factorization of n is of the form

$$n = p^a q_1^{2b_1} q_2^{2b_2} \cdots q_t^{2b_t},$$

where both a and p are congruent to 1 modulo 4 and $t \geq 2$.

Proof: Let

$$n = p_1^{a_1} p_2^{a_2} \cdots p_k^{a_k}$$

be the canonical prime factorization of n. Since n is perfect, we have

$$\prod_{i=1}^{k}(1 + p_i + p_i^2 + \cdots + p_i^{a_i}) = 2p_1^{a_1} p_2^{a_2} \cdots p_k^{a_k}.$$

Since n is odd, there is exactly one i, $1 \leq i \leq k$, such that

$$1 + p_i + p_i^2 + \cdots + p_i^{a_i} \equiv 2 \pmod 4.$$

Then a_i must be odd. Write $a_i = 2x + 1$ for some integer x. Since $p_i^2 \equiv 1 \pmod 4$, we can rewrite the above congruence equation as $(x + 1)(p_i + 1) \equiv 2 \pmod 4$, implying that $p_i \equiv 1 \pmod 4$ and x is even, and so $a_i \equiv 1 \pmod 4$.

For $j \neq i$ with $1 \leq j \leq k$, we have

$$1 + p_j + p_j^2 + \cdots + p_j^{a_j} \equiv 1 \pmod 2,$$

and so j must be even. It follows that

$$n = p^a q_1^{2b_1} q_2^{2b_2} \cdots q_t^{2b_t},$$

where both a and p are congruent to 1 modulo 4.

It remains to show that $t \geq 2$. Assume to the contrary that $t = 1$. We have

$$(1 + p + p^2 + \cdots + p_1^a)(1 + q + q^2 + \cdots + p_2^{2b}) = 2p^a q^{2b},$$

or

$$\frac{p^{a+1} - 1}{p - 1} \cdot \frac{q^{2b+1} - 1}{q - 1} = 2p^a q^{2b}.$$

It follows that

$$2 = \frac{p - \frac{1}{p^a}}{p - 1} \cdot \frac{q - \frac{1}{q^{2b}}}{q - 1} < \frac{p}{p - 1} \cdot \frac{q}{q - 1} \leq \frac{5}{4} \cdot \frac{3}{2} = \frac{15}{8},$$

which is not true. Hence our assumption was wrong and $t \geq 2$. □

In 1980, Hagis proved that $t \geq 7$ and $n > 10^{50}$. The existence of odd perfect numbers still remains one of the most challenging problems in number theory.

2

Introductory Problems

1. Let $1, 4, \ldots$ and $9, 16, \ldots$ be two arithmetic progressions. The set S is the union of the first 2004 terms of each sequence. How many distinct numbers are in S?

2. Given a sequence of six strictly increasing positive integers such that each number (besides the first) is a multiple of the one before it and the sum of all six numbers is 79, what is the largest number in the sequence?

3. What is the largest positive integer n for which $n^3 + 100$ is divisible by $n + 10$?

4. Those irreducible fractions!

   (1) Let n be an integer greater than 2. Prove that among the fractions

   $$\frac{1}{n}, \frac{2}{n}, \ldots, \frac{n-1}{n},$$

   an even number are irreducible.

   (2) Show that the fraction

   $$\frac{12n+1}{30n+2}$$

   is irreducible for all positive integers n.

5. A positive integer is written on each face of a cube. Each vertex is then assigned the product of the numbers written on the three faces intersecting the vertex. The sum of the numbers assigned to all the vertices is equal to 1001. Find the sum of the numbers written on the faces of the cube.

6. Call a number *prime looking* if it is composite but not divisible by 2, 3, or 5. The three smallest prime-looking numbers are 49, 77, and 91. There are 168 prime numbers less than 1000. How many prime-looking numbers are there less than 1000?

7. A positive integer k greater than 1 is given. Prove that there exist a prime p and a strictly increasing sequence of positive integers $a_1, a_2, \dots, a_n, \dots$ such that the terms of the sequence

$$p + ka_1, p + ka_2, \dots, p + ka_n, \dots$$

are all primes.

8. Given a positive integer n, let $p(n)$ be the product of the nonzero digits of n. (If n has only one digit, then $p(n)$ is equal to that digit.) Let

$$S = p(1) + p(2) + \cdots + p(999).$$

What is the largest prime factor of S?

9. Let m and n be positive integers such that

$$\operatorname{lcm}(m, n) + \gcd(m, n) = m + n.$$

Prove that one of the two numbers is divisible by the other.

10. Let $n = 2^{31}3^{19}$. How many positive integer divisors of n^2 are less than n but do not divide n?

11. Show that for any positive integers a and b, the number

$$(36a + b)(a + 36b)$$

cannot be a power of 2.

12. Compute the sum of the greatest odd divisor of each of the numbers 2006, 2007, ..., 4012.

13. Compute the sum of all numbers of the form a/b, where a and b are relatively prime positive divisors of 27000.

14. L.C.M of three numbers.

    (1) Find the number of ordered triples (a, b, c) of positive integers for which $\operatorname{lcm}(a, b) = 1000$, $\operatorname{lcm}(b, c) = 2000$, and $\operatorname{lcm}(c, a) = 2000$.

    (2) Let a, b, and c be integers. Prove that

$$\frac{\operatorname{lcm}(a, b, c)^2}{\operatorname{lcm}(a, b)\operatorname{lcm}(b, c)\operatorname{lcm}(c, a)} = \frac{\gcd(a, b, c)^2}{\gcd(a, b)\gcd(b, c)\gcd(c, a)}.$$

15. Let x, y, z be positive integers such that

$$\frac{1}{x} - \frac{1}{y} = \frac{1}{z}.$$

    Let h be the greatest common divisor of x, y, z. Prove that $hxyz$ and $h(y - x)$ are perfect squares.

16. Let p be a prime of the form $3k + 2$ that divides $a^2 + ab + b^2$ for some integers a and b. Prove that a and b are both divisible by p.

17. The number 27000001 has exactly four prime factors. Find their sum.

18. Find all positive integers n for which $n! + 5$ is a perfect cube.

19. Find all primes p such that the number $p^2 + 11$ has exactly six different divisors (including 1 and the number itself).

20. Call a positive integer N a *7-10 double* if the digits of the base-7 representation of N form a base-10 number that is twice N. For example, 51 is a 7-10 double because its base-7 representation is 102. What is the largest 7-10 double?

21. If $a \equiv b \pmod{n}$, show that $a^n \equiv b^n \pmod{n^2}$. Is the converse true?

22. Let p be a prime, and let $1 \le k \le p - 1$ be an integer. Prove that

$$\binom{p-1}{k} \equiv (-1)^k \pmod{p}.$$

23. Let p be a prime. Show that there are infinitely many positive integers n such that p divides $2^n - n$.

24. Let n be an integer greater than three. Prove that $1! + 2! + \cdots + n!$ cannot be a perfect power.

25. Let k be an odd positive integer. Prove that

$$(1 + 2 + \cdots + n) \mid (1^k + 2^k + \cdots + n^k)$$

for all positive integers n.

26. Let p be a prime greater than 5. Prove that $p - 4$ cannot be the fourth power of an integer.

27. For a positive integer n, prove that

$$\sigma(1) + \sigma(2) + \cdots + \sigma(n) \leq n^2.$$

28. Determine all finite nonempty sets S of positive integers satisfying

$$\frac{i + j}{\gcd(i, j)}$$

is an element of S for all i and j (not necessarily distinct) in S.

29. Knowing that 2^{29} is a nine-digit number all of whose digits are distinct, without computing the actual number determine which of the ten digits is missing. Justify your answer.

30. Prove that for any integer n greater than 1, the number $n^5 + n^4 + 1$ is composite.

31. The product of a few primes is ten times as much as the sum of the primes. What are these (not necessarily distinct) primes?

32. A 10-digit number is said to be *interesting* if its digits are all distinct and it is a multiple of 11111. How many interesting integers are there?

33. Do there exist 19 distinct positive integers that add up to 1999 and have the same sum of digits?

34. Find all prime numbers p and q such that pq divides the product $(5^p - 2^p)(5^q - 2^q)$.

35. Prove that there are infinitely many numbers not containing the digit 0 that are divisible by the sum of their digits.

36. Prove that any number consisting of 2^n identical digits has at least n distinct prime factors.

37. Let a and b be two relatively prime positive integers, and consider the arithmetic progression $a, a + b, a + 2b, a + 3b, \ldots$.

    (1) Prove that there are infinitely many terms in the arithmetic progression that have the same prime divisors.

    (2) Prove that there are infinitely many pairwise relatively prime terms in the arithmetic progression.

38. Let n be a positive integer.

    (1) Evaluate $\gcd(n! + 1, (n + 1)! + 1)$.

    (2) Let a and b be positive integers. Prove that

    $$\gcd(n^a - 1, n^b - 1) = n^{\gcd(a,b)} - 1.$$

    (3) Let a and b be positive integers. Prove that $\gcd(n^a + 1, n^b + 1)$ divides $n^{\gcd(a,b)} + 1$.

    (4) Let m be a positive integer with $\gcd(m, n) = 1$. Express

    $$\gcd(5^m + 7^m, 5^n + 7^n)$$

    in terms of m and n.

39. Bases? What bases?

(1) Determine whether it is possible to find a cube and a plane such that the distances from the vertices of the cube to the plane are 0, 1, 2, ..., 7.

(2) The increasing sequence 1, 3, 4, 9, 10, 12, 13, ... consists of all those positive integers that are powers of 3 or sums of distinct powers of 3. Find the 100th term of this sequence (where 1 is the 1st term, 3 is the 2nd term, and so on).

40. Fractions in modular arithmetic.

(1) Let a be the integer such that

$$1 + \frac{1}{2} + \frac{1}{3} + \cdots + \frac{1}{23} = \frac{a}{23!}.$$

Compute the remainder when a is divided by 13.

(2) Let $p > 3$ be a prime, and let m and n be relatively prime integers such that

$$\frac{m}{n} = \frac{1}{1^2} + \frac{1}{2^2} + \cdots + \frac{1}{(p-1)^2}.$$

Prove that m is divisible by p.

(3) Let $p > 3$ be a prime. Prove that Let $p > 3$ be a prime. Prove that

$$p^2 \mid (p-1)!\left(1 + \frac{1}{2} + \cdots + \frac{1}{p-1}\right).$$

41. Find all pairs (x, y) of nonnegative integers such that $x^2 + 3y$ and $y^2 + 3x$ are simultaneously perfect squares.

42. First digit? Not the last digit? Are your sure?

(1) Given that 2^{2004} is a 604-digit number with leading digit 1, determine the number of elements in the set

$$\{2^0,\ 2^1,\ 2^2,\ \ldots,\ 2^{2003}\}$$

with leading digit 4.

(2) Let k be a positive integer and let $n = n(k)$ be a positive integer such that in decimal representation 2^n and 5^n begin with the same k digits. What are these digits?

43. What are those missing digits?

(1) Determine the respective last digit (unit digit) of the numbers

$$3^{1001}7^{1002}13^{1003} \quad \text{and} \quad \underbrace{7^{7^{7^{\cdot^{\cdot^{\cdot^{7}}}}}}}_{1001\ 7\text{'s}}.$$

(2) Determine the last three digits of the number

$$2003^{2002^{2001}}.$$

(3) The binomial coefficient $\binom{99}{19}$ is a 21-digit number:

$$107{,}196{,}674{,}080{,}761{,}936{,}xyz.$$

Find the three-digit number xyz.

(4) Find the smallest positive integer whose cube ends in 888.

44. Let $p \geq 3$ be a prime, and let

$$\{a_1, a_2, \ldots, a_{p-1}\} \quad \text{and} \quad \{b_1, b_2, \ldots, b_{p-1}\}$$

be two sets of complete residue classes modulo p. Prove that

$$\{a_1b_1, a_2b_2, \ldots, a_{p-1}b_{p-1}\}$$

is not a complete set of residue classes modulo p.

45. Let $p \geq 3$ be a prime. Determine whether there exists a permutation

$$(a_1, a_2, \ldots, a_{p-1})$$

of $(1, 2, \ldots, p-1)$ such that the sequence $\{ia_i\}_{i=1}^{p-1}$ contains $p-2$ distinct congruence classes modulo p.

46. Prove that any positive integer less than $n!$ can be represented as a sum of no more than n positive integer divisors of $n!$.

47. Let $n > 1$ be an odd integer. Prove that n does not divide $3^n + 1$.

48. Let a and b be positive integers. Prove that the number of solutions (x, y, z) in nonnegative integers to the equation $ax + by + z = ab$ is

$$\frac{1}{2}[(a+1)(b+1) + \gcd(a, b) + 1].$$

49. Order! Order, please!

    (1) Let p be an odd prime, and let q and r be primes such that p divides $q^r + 1$. Prove that either $2r \mid p - 1$ or $p \mid q^2 - 1$.

    (2) Let $a > 1$ and n be given positive integers. If p is a prime divisor of $a^{2^n} + 1$, prove that $p - 1$ is divisible by 2^{n+1}.

50. Prove that

$$\left\lfloor \frac{(n-1)!}{n(n+1)} \right\rfloor$$

is even for every positive integer n.

51. Determine all the positive integers m each of which satisfies the following property: there exists a unique positive integer n such that there exist rectangles that can be divided into n congruent squares and also into $n + m$ congruent squares.

52. Determine all positive integers n such that n has a multiple whose digits are nonzero.

3

Advanced Problems

1. (a) Prove that the sum of the squares of 3, 4, 5, or 6 consecutive integers is not a perfect square.

   (b) Give an example of 11 consecutive positive integers the sum of whose squares is a perfect square.

2. Let $S(x)$ be the sum of the digits of the positive integer x in its decimal representation.

   (a) Prove that for every positive integer x, $\frac{S(x)}{S(2x)} \leq 5$. Can this bound be improved?

   (b) Prove that $\frac{S(x)}{S(3x)}$ is not bounded.

3. Most positive integers can be expressed as a sum of two or more consecutive positive integers. For example, $24 = 7 + 8 + 9$ and $51 = 25 + 26$. A positive integer that cannot be expressed as a sum of two or more consecutive positive integers is therefore *interesting*. What are all the interesting integers?

4. Set $S = \{105, 106, \ldots, 210\}$. Determine the minimum value of n such that any n-element subset T of S contains at least two non-relatively prime elements.

5. The number

$$\underbrace{99\ldots99}_{1997\ 9\text{'s}}$$

is written on a blackboard. Each minute, one number written on the blackboard is factored into two factors and erased, each factor is (independently) increased or diminished by 2, and the resulting two numbers are written. Is it possible that at some point (after the first minute) all of the numbers on the blackboard equal 9?

6. Let d be any positive integer not equal to 2, 5, or 13. Show that one can find distinct a, b in the set $\{2, 5, 13, d\}$ such that $ab - 1$ is not a perfect square.

7. A heap of balls consists of one thousand 10-gram balls and one thousand 9.9-gram balls. We wish to pick out two heaps of balls with equal numbers of balls in them but different total weights. What is the minimal number of weighings needed to do this? (The balance scale reports the weight of the objects in the left pan minus the weight of the objects in the right pan.)

8. We are given three integers a, b, and c such that $a, b, c, a+b-c, a+c-b, b+c-a$, and $a+b+c$ are seven distinct primes. Let d be the difference between the largest and smallest of these seven primes. Suppose that 800 is an element in the set $\{a+b, b+c, c+a\}$. Determine the maximum possible value of d.

9. Prove that the sum

$$S(m, n) = \frac{1}{m} + \frac{1}{m+1} + \cdots + \frac{1}{m+n}$$

is not an integer for any given positive integers m and n.

10. For all positive integers $m > n$, prove that

$$\operatorname{lcm}(m, n) + \operatorname{lcm}(m+1, n+1) > \frac{2mn}{\sqrt{m-n}}.$$

11. Prove that each nonnegative integer can be represented in the form $a^2 + b^2 - c^2$, where a, b, and c are positive integers with $a < b < c$.

12. Determine whether there exists a sequence of strictly increasing positive integers $\{a_k\}_{k=1}^{\infty}$ such that the sequence $\{a_k + a\}_{k=1}^{\infty}$ contains only finitely many primes for all integers a.

13. Prove that for different choices of signs $+$ and $-$ the expression

$$\pm 1 \pm 2 \pm 3 \pm \cdots \pm (4n+1)$$

yields all odd positive integers less than or equal to $(2n+1)(4n+1)$.

14. Let a and b be relatively prime positive integers. Show that

$$ax + by = n$$

has nonnegative integer solutions (x, y) for all integers $n > ab - a - b$. What if $n = ab - a - b$?

15. The sides of a triangle have integer lengths k, m, and n. Assume that $k > m > n$ and

$$\left\{ \frac{3^k}{10^4} \right\} = \left\{ \frac{3^m}{10^4} \right\} = \left\{ \frac{3^n}{10^4} \right\}.$$

Determine the minimum value of the perimeter of the triangle.

16. Consider the following two-person game. A number of pebbles are lying on a table. Two players make their moves alternately. A move consists in taking off the table x pebbles, where x is the square of any positive integer. The player who is unable to make a move loses. Prove that there are infinitely many initial situations in which the player who goes second has a winning strategy?

17. Prove that the sequence $1, 11, 111, \ldots$ contains an infinite subsequence whose terms are pairwise relatively prime.

18. Let m and n be integers greater than 1 such that $\gcd(m, n-1) = \gcd(m, n) = 1$. Prove that the first $m-1$ terms of the sequence $n_1, n_2, \ldots,$ where $n_1 = mn + 1$ and $n_{k+1} = n \cdot n_k + 1$, $k \geq 1$, cannot all be primes.

19. Find all positive integers m such that the fourth power of the number of positive divisors of m equals m.

20. (1) Show that it is possible to choose one number out of any 39 consecutive positive integers having the sum of its digits divisible by 11.
    (2) Find the first 38 consecutive positive integers none of which has the sum of its digits divisible by 11.

21. Find the largest integer n such that n is divisible by all positive integers less than $\sqrt[3]{n}$.

22. Show that for any fixed positive integer n, the sequence

$$2,\ 2^2,\ 2^{2^2},\ 2^{2^{2^2}},\ \dots \pmod{n}$$

is eventually constant. (The tower of exponents is defined by $a_1 = 2$ and $a_{i+1} = 2^{a_i}$ for every positive integer i.)

23. Prove that for $n \geq 5$, $f_n + f_{n-1} - 1$ has at least $n + 1$ prime factors, where $f_n = 2^{2^n} + 1$.

24. Prove that any integer can be written as the sum of the cubes of five integers, not necessarily distinct.

25. Integer or fractional parts?
    (1) Find all real numbers x such that

$$x\lfloor x\lfloor x\lfloor x\rfloor\rfloor\rfloor = 88.$$

    (2) Show that the equation

$$\{x^3\} + \{y^3\} = \{z^3\}$$

has infinitely many rational noninteger solutions.

26. Let n be a given positive integer. If p is a prime divisor of the Fermat number f_n, prove that $p - 1$ is divisible by 2^{n+2}.

27. The sequence

$$\{a_n\}_{n=1}^{\infty} = \{1, 2, 4, 5, 7, 9, 10, 12, 14, 16, 17, \dots\}$$

of positive integers is formed by taking one odd integer, then two even integers, then three odd integers, etc. Express a_n in closed form.

28. Prove that for each $n \geq 2$, there is a set S of n integers such that $(a-b)^2$ divides ab for every distinct $a, b \in S$.

29. Show that there exist infinitely many positive integers n such that the largest prime divisor of n^4+1 is greater than $2n$.

30. For a positive integer k, let $p(k)$ denote the greatest odd divisor of k. Prove that for every positive integer n,

$$\frac{2n}{3} < \frac{p(1)}{1} + \frac{p(2)}{2} + \cdots + \frac{p(n)}{n} < \frac{2(n+1)}{3}.$$

31. If p^t is an odd prime power and m is an integer relatively prime to both p and $p-1$, then for any a and b relatively prime to p,

$$a^m \equiv b^m \pmod{p^t} \text{ if and only if } a \equiv b \pmod{p^t}.$$

32. Prove that for each prime $p \geq 7$, there exists a positive integer n and integers $x_1, \ldots, x_n, y_1, \ldots, y_n$ not divisible by p such that

$$\begin{cases} x_1^2 + y_1^2 \equiv x_2^2 \pmod{p}, \\ x_2^2 + y_2^2 \equiv x_3^2 \pmod{p}, \\ \vdots \\ x_n^2 + y_n^2 \equiv x_1^2 \pmod{p}. \end{cases}$$

33. For every positive integer n, prove that

$$\frac{\sigma(1)}{1} + \frac{\sigma(2)}{2} + \cdots + \frac{\sigma(n)}{n} \leq 2n.$$

34. Prove that the system

$$\begin{aligned} x^6 + x^3 + x^3y + y &= 147^{157}, \\ x^3 + x^3y + y^2 + y + z^9 &= 157^{147}, \end{aligned}$$

has no solutions in integers x, y, and z.

35. What is the smallest number of weighings on a balance scale needed to identify the individual weights of a set of objects known to weigh $1, 3, 3^2, \ldots, 3^{26}$ in some order? (The balance scale reports the weight of the objects in the left pan minus the weight of the objects in the right pan.)

36. Let λ be the positive root of the equation $t^2 - 1998t - 1 = 0$. Define the sequence $x_0, x_1, \ldots$ by setting

$$x_0 = 1, \quad x_{n+1} = \lfloor \lambda x_n \rfloor \qquad (n \geq 0).$$

Find the remainder when x_{1998} is divided by 1998.

37. Determine (with proof) whether there is a subset X of the integers with the following property: for any integer n there is exactly one solution of $a + 2b = n$ with $a, b \in X$.

38. The number x_n is defined as the last digit in the decimal representation of the integer $\left\lfloor \sqrt{2}^n \right\rfloor$ $(n = 1, 2, \ldots)$. Determine whether the sequence $x_1, x_2, \ldots, x_n, \ldots$ is periodic.

39. Prove that every integer n can be represented in infinitely many ways as

$$n = \pm 1^2 \pm 2^2 \pm \cdots \pm k^2$$

for a convenient k and a suitable choice of the signs $+$ and $-$.

40. Let n be a given integer with $n \geq 4$. For a positive integer m, let S_m denote the set $\{m, m+1, \ldots, m+n-1\}$. Determine the minimum value of $f(n)$ such that every $f(n)$-element subset of S_m (for every m) contains at least three pairwise relatively prime elements.

41. Find the least positive integer r such that for any positive integers a, b, c, d, $((abcd)!)^r$ is divisible by the product of

$$\begin{gathered}(a!)^{bcd+1}, \quad (b!)^{acd+1}, \quad (c!)^{abd+1}, \quad (d!)^{abc+1},\\ ((ab)!)^{cd+1}, \ ((bc)!)^{ad+1}, \ ((cd)!)^{ab+1}, \ ((ac)!)^{bd+1},\\ ((bd)!)^{ac+1}, \ ((ad)!)^{bc+1}, \ ((abc)!)^{d+1}, \ ((abd)!)^{c+1},\\ ((acd)!)^{b+1}, \ ((bcd)!)^{a+1}.\end{gathered}$$

42. Two classics on L.C.M.

(1) Let $a_0 < a_1 < a_2 < \cdots < a_n$ be positive integers. Prove that

$$\frac{1}{\operatorname{lcm}(a_0, a_1)} + \frac{1}{\operatorname{lcm}(a_1, a_2)} + \cdots + \frac{1}{\operatorname{lcm}(a_{n-1}, a_n)} \le 1 - \frac{1}{2^n}.$$

(2) Several positive integers are given not exceeding a fixed integer constant m. Prove that if every positive integer less than or equal to m is not divisible by any pair of the given numbers, then the sum of the reciprocals of these numbers is less than $\frac{3}{2}$.

43. For a positive integer n, let $r(n)$ denote the sum of the remainders of n divided by $1, 2, \ldots, n$. Prove that there are infinitely many n such that $r(n) = r(n-1)$.

44. Two related IMO problems.

(1) A *wobbly number* is a positive integer whose digits are alternately nonzero and zero with the units digit being nonzero. Determine all positive integers that do not divide any wobbly numbers.

(2) A positive integer is called *alternating* if among any two consecutive digits in its decimal representation, one is even and the other is odd. Find all positive integers n such that n has a multiple that is alternating.

45. Let p be an odd prime. The sequence $(a_n)_{n \ge 0}$ is defined as follows: $a_0 = 0$, $a_1 = 1, \ldots, a_{p-2} = p - 2$, and for all $n \ge p - 1$, a_n is the least positive integer that does not form an arithmetic sequence of length p with any of the preceding terms. Prove that for all n, a_n is the number obtained by writing n in base $p - 1$ and reading the result in base p.

46. Determine whether there exists a positive integer n such that n is divisible by exactly 2000 different prime numbers, and $2^n + 1$ is divisible by n.

47. Two cyclic symmetric divisibility relations.

(1) [Russia 2000] Determine whether there exist pairwise relatively prime integers a, b, and c with $a, b, c > 1$ such that

$$b \mid 2^a + 1, \quad c \mid 2^b + 1, \quad a \mid 2^c + 1.$$

(2) [TST 2003, by Reid Barton] Find all ordered triples of primes (p, q, r) such that

$$p \mid q^r + 1, \quad q \mid r^p + 1, \quad r \mid p^q + 1.$$

48. Let n be a positive integer, and let $p_1, p_2, \ldots, p_n$ be distinct primes greater than 3. Prove that $2^{p_1 p_2 \cdots p_n} + 1$ has at least 4^n divisors.

49. Let p be a prime, and let $\{a_k\}_{k=0}^{\infty}$ be a sequence of integers such that $a_0 = 0$, $a_1 = 1$, and

$$a_{k+2} = 2a_{k+1} - pa_k$$

for $k = 0, 1, 2, \ldots$. Suppose that -1 appears in the sequence. Find all possible values of p.

50. Let $\mathcal{F}$ be a set of subsets of the set $\{1, 2, \ldots, n\}$ such that

(1) if A is an element of $\mathcal{F}$, then A contains exactly three elements;

(2) if A and B are two distinct elements in $\mathcal{F}$, A and B share at most one common element.

Let $f(n)$ denote the maximum number of elements in $\mathcal{F}$. Prove that

$$\frac{(n-1)(n-2)}{6} \le f(n) \le \frac{(n-1)n}{6}.$$

51. Determine all positive integers k such that

$$\frac{\tau(n^2)}{\tau(n)} = k,$$

for some n.

52. Let n be a positive integer greater than two. Prove that the Fermat number f_n has a prime divisor greater than $2^{n+2}(n+1)$.

4

Solutions to Introductory Problems

1. [AMC10B 2004] Let $1, 4, \ldots$ and $9, 16, \ldots$ be two arithmetic progressions. The set S is the union of the first 2004 terms of each sequence. How many distinct numbers are in S?

   Solution: The smallest number that appears in both sequences is 16. Since the least common multiple of 3 and 7 (the two common differences of the progressions) is 21, numbers appear in both sequences only if they are in the form $16+21k$, where k is a nonnegative integer. The largest k such that $7k + 9 \le 2004$ is $k = 285$. Hence there are 286 numbers each of which appears in both progressions. Thus the answer is $4008 - 286 = 3722$.

2. [HMMT 2004] Given a sequence of six strictly increasing positive integers such that each number (besides the first) is a multiple of the one before it and the sum of all six numbers is 79, what is the largest number in the sequence?

   Solution: Let $a_1 < a_2 < \cdots < a_6$ be the six numbers. If $a_4 \ge 12$, then $a_5 \ge 2a_4 \ge 24$ and $a_6 \ge 2a_5 \ge 48$, implying that $a_4+a_5+a_6 \ge 84$, which violates the conditions of the problem. Hence $a_4 < 12$. Then the only way we can have the required divisibilities among the first four numbers is if they are $a_1 = 1$, $a_2 = 2$, $a_3 = 4$, and $a_4 = 8$. We write $a_5 = ma_4 = 8m$ and $a_6 = na_5 = 8mn$ for integers m and n with $m, n \ge 2$. We get $8m + 8mn = 79 - (1 + 2 + 4 + 8) = 64$, or $m(1 + n) = 8$. This leads to the unique solution $m = 2$ and $n = 3$. Hence the answer is $a_6 = 48$.

3. [AIME 1986] What is the largest positive integer n for which $n^3 + 100$ is divisible by $n + 10$?

Solution: By division we find that $n^3 + 100 = (n + 10)(n^2 - 10n + 100) - 900$. Thus, if $n + 10$ divides $n^3 + 100$, then it must also divide 900. Moreover, since n is maximized whenever $n + 10$ is, and since the largest divisor of 900 is 900, we must have $n + 10 = 900$. Therefore, $n = 890$.

4. Those irreducible fractions!

(1) Let n be an integer greater than 2. Prove that among the fractions

$$\frac{1}{n}, \frac{2}{n}, \ldots, \frac{n-1}{n},$$

an even number of them are irreducible.

(2) Show that the fraction

$$\frac{12n+1}{30n+2}$$

is irreducible for all positive integers n.

Proof: We prove part (1) via a parity argument, and we establish part (2) applying the Euclidean algorithm.

(1) The fraction $\frac{k}{n}$ is irreducible if and only if the fraction $\frac{n-k}{n}$ is irreducible, because $\gcd(k, n) = \gcd(n - k, n)$.
If the fractions $\frac{k}{n}$ and $\frac{n-k}{n}$ are distinct for all k, then pairing up yields an even number of irreducible fractions.
If $\frac{k}{n} = \frac{n-k}{n}$ for some k, then $n = 2k$ and so $\frac{k}{n} = \frac{k}{2k} = \frac{1}{2}$ is reducible and the problem reduces to the previous case.

(2) Note that

$$\gcd(30n + 2, 12n + 1) = \gcd(6n, 12n + 1) = \gcd(6n, 1) = 1,$$

from which the desired result follows.

5. A positive integer is written on each face of a cube. Each vertex is then assigned the product of the numbers written on the three faces intersecting the vertex. The sum of the numbers assigned to all the vertices is equal to 1001. Find the sum of the numbers written on the faces of the cube.

Solution: Let a, b, c, d, e, and f be the numbers written on the faces, with a and f, b and d, c and e written on opposite faces. We are given that

$$\begin{aligned} 1001 &= abc + abe + acd + ade + bcf + bef + cdf + def \\ &= (a + f)(b + d)(c + e). \end{aligned}$$

(We can realize this factorization by noticing that the product xyz appears exactly once if and only if x and y, y and z, z and x are not written on the opposite faces.) Since $1001 = 7 \cdot 11 \cdot 13$ and each of $a + f$, $b + d$, and $c + e$ are greater than 1, it follows that $\{a + f, b + d, c + e\} = \{7, 11, 13\}$, implying that the answer is $a + b + c + d + e + f = 7 + 11 + 13 = 31$.

6. [AMC12A 2005] Call a number *prime looking* if it is composite but not divisible by 2, 3, or 5. The three smallest prime-looking numbers are 49, 77, and 91. There are 168 prime numbers less than 1000. How many prime-looking numbers are there less than 1000?

Solution: Of the numbers less than 1000, $\left\lfloor \frac{999}{2} \right\rfloor = 499$ of them are divisible by two, $\left\lfloor \frac{999}{3} \right\rfloor = 333$ are divisible by 3, and $\left\lfloor \frac{999}{5} \right\rfloor = 199$ are divisible by 5. There are $\left\lfloor \frac{999}{6} \right\rfloor = 166$ multiples of 6, $\left\lfloor \frac{999}{10} \right\rfloor = 99$ multiples of 10, and $\left\lfloor \frac{999}{15} \right\rfloor = 66$ multiples of 15. Finally, there are $\left\lfloor \frac{999}{30} \right\rfloor = 33$ multiples of 30. By the inclusion and exclusion principle there are

$$499 + 333 + 199 - 166 - 99 - 66 + 33 = 733$$

numbers that are divisible by at least one of 2, 3, and 5. Of the remaining $999 - 733 = 266$ numbers, 165 are primes other than 2, 3, or 5. Note that 1 is neither prime nor composite. This leaves exactly 100 prime-looking numbers.

7. A positive integer k greater than 1 is given. Prove that there exist a prime p and a strictly increasing sequence of positive integers $a_1, a_2, \ldots, a_n, \ldots$ such that the terms of the sequence

$$p + ka_1, p + ka_2, \ldots, p + ka_n, \ldots$$

are all primes.

Proof: The pigeonhole principle provides an elegant solution. There is nothing to be afraid of, just infinitely many pigeons in finitely many pigeonholes.

For each $i = 1, 2, \ldots, k-1$ denote by P_i the set of all primes congruent to i modulo k. Each prime (except possibly k itself) is contained in exactly one of the sets $P_1, P_2, \ldots, P_{k-1}$. Because there are infinitely many primes, at least one of these sets is infinite, say P_i. Let $p = x_1 < x_2 < \cdots < x_n < \cdots$ be its elements arranged in increasing order, and

$$a_n = \frac{x_{n+1} - p}{k}$$

for every positive integer n. Then the $p + ka_n$ simply run through the members of P_i, beginning at x_2. The numbers a_n are positive integers. The prime p and the strictly increasing sequence $a_1, a_2, \ldots, a_n, \ldots$ have the desired properties.

8. [AIME 1994] Given a positive integer n, let $p(n)$ be the product of the nonzero digits of n. (If n has only one digit, then $p(n)$ is equal to that digit.) Let

$$S = p(1) + p(2) + \cdots + p(999).$$

What is the largest prime factor of S?

Solution: Consider each positive integer less than 1000 to be a three-digit number by prefixing 0's to numbers with fewer than three digits. The sum of the products of the digits of all such positive numbers is

$$(0 \cdot 0 \cdot 0 + 0 \cdot 0 \cdot 1 + \cdots + 9 \cdot 9 \cdot 9) - 0 \cdot 0 \cdot 0 = (0 + 1 + \cdots + 9)^3 - 0.$$

However, $p(n)$ is the product of nonzero digits of n. The sum of these products can be found by replacing 0 by 1 in the above expression, since ignoring 0's is equivalent to thinking of them as 1's in the products. (Note that the final 0 in the above expression becomes a 1 and compensates for the contribution of 000 after it is changed to 111.) Hence

$$S = 46^3 - 1 = (46 - 1)(46^2 + 46 + 1) = 3^3 \cdot 5 \cdot 7 \cdot 103,$$

and the largest prime factor is 103.

9. [Russia 1995] Let m and n be positive integers such that

$$\operatorname{lcm}(m, n) + \gcd(m, n) = m + n.$$

Prove that one of the two numbers is divisible by the other.

First Proof: Let $d = \gcd(m, n)$. We write $m = ad$ and $n = bd$. Then $\gcd(a, b) = 1$ and

$$\operatorname{lcm}(m, n) = \frac{mn}{\gcd(m, n)} = abd.$$

The given equation becomes $abd + d = ad + bd$, or $ab - a - b + 1 = 0$. It follows that $(a - 1)(b - 1) = 0$, implying that either $a = 1$ or $b = 1$; that is, either $m = d$, $n = bd = bm$ or $n = d$, $m = an$.

Second Proof: Because $\operatorname{lcm}(m, n) \cdot \gcd(m, n) = mn$, it follows that $\operatorname{lcm}(m, n)$ and $\gcd(m, n)$ as well as m, n are roots of $x^2 - (m + n)x + mn = 0$. Hence $\{\operatorname{lcm}(m, n), \gcd(m, n)\} = \{m, n\}$ and the conclusion follows.

10. [AIME 1995] Let $n = 2^{31}3^{19}$. How many positive integer divisors of n^2 are less than n but do not divide n?

First Solution: Let $n = p^r q^s$, where p and q are distinct primes. Then $n^2 = p^{2r} q^{2s}$, so n^2 has

$$(2r + 1)(2s + 1)$$

factors. For each factor less than n, there is a corresponding factor greater than n. By excluding the factor n, we see that there must be

$$\frac{(2r + 1)(2s + 1) - 1}{2} = 2rs + r + s$$

factors of n^2 that are less than n. Because n has $(r + 1)(s + 1)$ factors (including n itself), and because every factor of n is also a factor of n^2, there are

$$2rs + r + s - [(r + 1)(s + 1) - 1] = rs$$

factors of n^2 that are less than n but not factors of n. When $r = 31$ and $s = 19$, there are $rs = 589$ such factors.

Second Solution: (By Chengde Feng) A positive integer divisor d of n^2 is less than n but does not divide n if and only if

$$d = \begin{cases} 2^{31+a}3^{19-b} & \text{if } 2^a < 3^b, \\ 2^{31-a}3^{19+b} & \text{if } 2^a > 3^b, \end{cases}$$

where a and b are integers such that $1 \le a \le 31$ and $1 \le b \le 19$. Since $2^a \ne 3^b$ for positive integers a and b, there are $19 \times 31 = 589$ such divisors.

11. [APMO 1998] Show that for any positive integers a and b, the number

$$(36a + b)(a + 36b)$$

cannot be a power of 2.

Proof: Write $a = 2^c \cdot p$, $b = 2^d \cdot q$, with p and q odd. Assume without loss of generality that $c \geq d$. Then

$$36a + b = 36 \cdot 2^c p + 2^d q = 2^d (36 \cdot 2^{c-d} p + q).$$

Consequently,

$$(36a + b)(36b + a) = 2^d (36 \cdot 2^{c-d} p + q)(36b + a)$$

has the nontrivial odd factor $36 \cdot 2^{c-d} p + q$, and thus is not a power of 2.

12. Compute sum of the greatest odd divisor of each of the numbers 2006, 2007, ..., 4012.

Solution: For a positive integer n, let $p(n)$ denote its greatest odd divisor. We can write $n = 2^k \cdot p(n)$ for some nonnegative integer k. If two positive integers n_1 and n_2 are such that $p(n_1) = p(n_2)$, then one is at least twice the other.

Because no number from 2007, 2008, ..., 4012 is twice another such number,
$p(2007)$, $p(2008)$, ..., $p(4012)$ are 2006 distinct odd positive integers. Also note that these odd numbers belong to the set $\{1, 3, 5, \ldots, 4011\}$, which also consists of exactly 2006 elements. It follows that

$$\{p(2007), p(2008), \ldots, p(4012)\} = \{1, 3, \ldots, 4011\}.$$

Hence the desired sum is equal to

$$\begin{aligned} p(2006) + 1 + 3 + \cdots + 4011 &= 1003 + 2006^2 \\ &= 1003 \cdot 4013 = 4025039. \end{aligned}$$

13. Compute the sum of all numbers of the form a/b, where a and b are relatively prime positive divisors of 27000.

Solution: Because $27000 = 2^3 3^3 5^3$, each a/b can be written in the form of $2^a 3^b 5^c$, where a, b, c are integers in the interval $[-3, 3]$. It follows that each a/b appears exactly once in the expansion of

$$(2^{-3} + 2^{-2} + \cdots + 2^3)(3^{-3} + 3^{-2} + \cdots + 3^3)(5^{-3} + 5^{-2} + \cdots + 5^3).$$

It follows that the desired sum is equal to

$$\frac{1}{2^3 3^3 5^3} \cdot \frac{2^7 - 1}{2 - 1} \cdot \frac{3^7 - 1}{3 - 1} \cdot \frac{5^7 - 1}{5 - 1} = \frac{(2^7 - 1)(3^7 - 1)(5^7 - 1)}{2^6 3^3 5^3}.$$

14. L.C.M of three numbers.

(1) [AIME 1987] Find the number of ordered triples (a, b, c) of positive integers for which $\operatorname{lcm}(a, b) = 1000$, $\operatorname{lcm}(b, c) = 2000$, and $\operatorname{lcm}(c, a) = 2000$.

(2) Let a, b, and c be integers. Prove that

$$\frac{\operatorname{lcm}(a, b, c)^2}{\operatorname{lcm}(a, b) \operatorname{lcm}(b, c) \operatorname{lcm}(c, a)} = \frac{\gcd(a, b, c)^2}{\gcd(a, b) \gcd(b, c) \gcd(c, a)}.$$

Solution: We have two different approaches to these two parts. For part (1), we deal with the L.C.M. and G.C.D. of three integers via pairwise L.C.M. and G.C.D. of two integers. For part (2), we use prime factorizations.

(1) Because both 1000 and 2000 are of the form $2^m 5^n$, the numbers a, b, and c must also be of this form. We set

$$a = 2^{m_1} 5^{n_1}, \quad b = 2^{m_2} 5^{n_2}, \qquad c = 2^{m_3} 5^{n_3},$$

where the m_i and n_i are nonnegative integers for $i = 1, 2, 3$. Then the following equalities must hold:

$$\max\{m_1, m_2\} = 3, \quad \max\{m_2, m_3\} = 4, \quad \max\{m_3, m_1\} = 4 \quad (*)$$

and

$$\max\{n_1, n_2\} = 3, \quad \max\{n_2, n_3\} = 3, \quad \max\{n_3, n_1\} = 3. \quad (**)$$

From $(*)$, we must have $m_3 = 4$, and either m_1 or m_2 must be 3, while the other one can take the values of 0, 1, 2, or 3. There are 7 such ordered triples, namely $(0, 3, 4)$, $(1, 3, 4)$, $(2, 3, 4)$, $(3, 0, 4)$, $(3, 1, 4)$, $(3, 2, 4)$, and $(3, 3, 4)$.

To satisfy (**), two of n_1, n_2, and n_3 must be 3, while the third one ranges through the values of 0, 1, 2, and 3. The number of such ordered triples is 10; they are (3, 3, 0), (3, 3, 1), (3, 3, 2), (3, 0, 3), (3, 1, 3), (3, 2, 3), (0, 3, 3), (1, 3, 3), (2, 3, 3), and (3, 3, 3). Because the choice of (m_1, m_2, m_3) is independent of the choice of (n_1, n_2, n_3), they can be chosen in $7 \cdot 10 = 70$ different ways. This is the number of ordered triples (a, b, c) satisfying the given conditions.

(2) Let $a = p_1^{\alpha_1} \cdots p_n^{\alpha_n}$, $b = p_1^{\beta_1} \cdots p_n^{\beta_n}$, and $c = p_1^{\gamma_1} \cdots p_n^{\gamma_n}$, where $p_1, \ldots, p_n$ are distinct primes, and $a_1, \ldots, a_n, b_1, \ldots, b_n, c_1, \ldots, c_n$ are nonzero integers. Then

$$\begin{aligned}&\frac{\operatorname{lcm}(a,b,c)^2}{\operatorname{lcm}(a,b)\operatorname{lcm}(b,c)\operatorname{lcm}(c,a)}\\&\qquad= \frac{\prod_{i=1}^n p_i^{2\max\{\alpha_i,\beta_i,\gamma_i\}}}{\prod_{i=1}^n p_i^{\max\{\alpha_i,\beta_i\}}\prod_{i=1}^n p_i^{\max\{\beta_i,\gamma_i\}}\prod_{i=1}^n p_i^{\max\{\gamma_i,\alpha_i\}}}\\&\qquad= \prod_{i=1}^n p_i^{2\max\{\alpha_i,\beta_i,\gamma_i\}-\max\{\alpha_i,\beta_i\}-\max\{\beta_i,\gamma_i\}-\max\{\gamma_i,\alpha_i\}}\end{aligned}$$

and

$$\begin{aligned}&\frac{\gcd(a,b,c)^2}{\gcd(a,b)\gcd(b,c)\gcd(c,a)}\\&\qquad= \frac{\prod_{i=1}^n p_i^{2\min\{\alpha_i,\beta_i,\gamma_i\}}}{\prod_{i=1}^n p_i^{\min\{\alpha_i,\beta_i\}}\prod_{i=1}^n p_i^{\min\{\beta_i,\gamma_i\}}\prod_{i=1}^n p_i^{\min\{\gamma_i,\alpha_i\}}}\\&\qquad= \prod_{i=1}^n p_i^{2\min\{\alpha_i,\beta_i,\gamma_i\}-\min\{\alpha_i,\beta_i\}-\min\{\beta_i,\gamma_i\}-\min\{\gamma_i,\alpha_i\}}\end{aligned}$$

It suffices to show that for each nonnegative numbers α, β, and γ,

$$\begin{aligned}&2\max\{\alpha,\beta,\gamma\} - \max\{\alpha,\beta\} - \max\{\beta,\gamma\} - \max\{\gamma,\alpha\}\\&\qquad= 2\min\{\alpha,\beta,\gamma\} - \min\{\alpha,\beta\} - \min\{\beta,\gamma\} - \min\{\gamma,\alpha\}.\end{aligned}$$

By symmetry, we may assume that $\alpha \le \beta \le \gamma$. It is not difficult to deduce that both sides are equal $-\beta$, completing our proof.

As a side result from the proof of (2), we note that

$$\frac{\operatorname{lcm}(a,b)\operatorname{lcm}(b,c)\operatorname{lcm}(c,a)}{\operatorname{lcm}(a,b,c)^2} \quad\text{and}\quad \frac{\gcd(a,b)\gcd(b,c)\gcd(c,a)}{\gcd(a,b,c)^2}$$

are equal integers.

15. [UK 1998] Let x, y, z be positive integers such that

$$\frac{1}{x} - \frac{1}{y} = \frac{1}{z}.$$

Let h be the greatest common divisor of x, y, z. Prove that $hxyz$ and $h(y - x)$ are perfect squares.

Proof: Let $x = ha$, $y = hb$, $z = hc$. Then a, b, c are positive integers such that $\gcd(a, b, c) = 1$. Let $\gcd(a, b) = g$. So $a = ga'$, $b = gb'$ and a' and b' are positive integers such that

$$\gcd(a', b') = \gcd(a' - b', b') = \gcd(a', a' - b') = 1.$$

We have

$$\frac{1}{a} - \frac{1}{b} = \frac{1}{c} \iff c(b - a) = ab \iff c(b' - a') = a'b'g.$$

So $g|c$ and $\gcd(a, b, c) = g = 1$. Therefore $\gcd(a, b) = 1$ and $\gcd(b - a, ab) = 1$. Thus $b - a = 1$ and $c = ab$. Now

$$hxyz = h^4abc = (h^2ab)^2 \quad \text{and} \quad h(y - x) = h^2$$

are both perfect squares, as desired.

16. Let p be a prime of the form $3k + 2$ that divides $a^2 + ab + b^2$ for some integers a and b. Prove that a and b are both divisible by p.

Proof: We approach indirectly by assuming that p does not divide a. Because p divides $a^2 + ab + b^2$, it also divides $a^3 - b^3 = (a - b)(a^2 + ab + b^2)$, so $a^3 \equiv b^3 \pmod{p}$. Hence

$$a^{3k} \equiv b^{3k} \pmod{p}.$$

Hence p does not divide b either. Applying Fermat's little theorem yields $a^{p-1} \equiv b^{p-1} \equiv 1 \pmod{p}$, or

$$a^{3k+1} \equiv b^{3k+1} \pmod{p}.$$

Because p is relatively prime to a, we conclude that $a \equiv b \pmod{p}$. This, combined with $a^2 + ab + b^2 \equiv 0 \pmod{p}$, implies $3a^2 \equiv 0 \pmod{p}$. Because $p \neq 3$, it turns out that p divides a, which is a contradiction.

17. [HMMT 2005] The number 27000001 has exactly four prime factors. Find their sum.

Solution: Since $x^3+1=(x+1)(x^2-x+1)$ and $x^2-y^2=(x+y)(x-y)$, it follows that

$$\begin{aligned}27000001 &= 300^3+1=(300+1)(300^2-300+1)\\ &= 301(300^2+2\cdot 300+1-900)\\ &= 301[(300+1)^2-900]=301(301^2-30^2)\\ &= 301\cdot 331\cdot 271=7\cdot 43\cdot 271\cdot 331.\end{aligned}$$

Hence the answer is $7+43+271+331=652$.

18. Find all positive integers n for which $n!+5$ is a perfect cube.

First Solution: The only answer is $n=5$.

One checks directly that $n!+5$ is not a perfect cube for $n=1,2,3,4,6,7,8,9$ and that $5!+5$ is a perfect cube.

If $n!+5$ were a perfect cube for $n>9$, then, since it is a multiple of 5, $n!+5$ would be a multiple of 125. However, this is not true, since $n!$ is a multiple of 125 for $n>9$, but 5 is not. Thus the only positive integer with the desired property is $n=5$.

Second Solution: Again, we check the cases $n=1,2,\ldots,6$ directly. For $n\geq 7$, $n!+5\equiv 5 \pmod 7$, which is not a cubic residue class modulo 7. (The only cubic residue classes modulo 7 are 0 and ± 1.)

19. [Russia 1995] Find all primes p such that the number p^2+11 has exactly six different divisors (including 1 and the number itself).

Solution: For $p\neq 3$, $p^2\equiv 1 \pmod 3$, and so $3\mid(p^2+11)$. Similarly, for $p\neq 2$, $p^2\equiv 1 \pmod 4$ and so $4\mid(p^2+11)$. Except in these two cases, then, $12\mid(p^2+11)$; since 12 itself has 6 divisors (1, 2, 3, 4, 6, 12) and $p^2+11>12$ for $p>1$, p^2+11 must have more than 6 divisors. The only cases to check are $p=2$ and $p=3$. If $p=2$, then $p^2+11=15$, which has only 4 divisors (1, 3, 5, 15), while if $p=3$, then $p^2+11=20$, which indeed has 6 divisors $\{1,2,4,5,10,20\}$. Hence $p=3$ is the only solution.

20. [AIME 2001] Call a positive integer N a *7-10 double* if the digits of the base-7 representation of N form a base-10 number that is twice N. For example, 51 is a 7-10 double because its base-7 representation is 102. What is the largest 7-10 double?

Solution: Suppose that $a_k 7^k + a_{k-1}7^{k-1} + \cdots + a_2 7^2 + a_1 7 + a_0$ is a 7-10 double, with $a_k \neq 0$. In other words, $a_k 10^k + a_{k-1}10^{k-1} + \cdots + a_2 10^2 + a_1 10 + a_0$ is twice as large, so that

$$a_k(10^k - 2 \cdot 7^k) + a_{k-1}(10^{k-1} - 2 \cdot 7^{k-1}) + \cdots + a_1(10 - 2 \cdot 7) + a_o(1 - 2) = 0.$$

Because the coefficient of a_i in this equation is negative only when $i = 0$ and $i = 1$, and no a_i is negative, it follows that k is at least 2. Because the coefficient of a_i is at least 314 when $i > 2$, and because no a_i exceeds 6, it follows that $k = 2$ and $2a_2 = 4a_1 + a_0$. To obtain the largest possible 7-10 double, first try $a_2 = 6$. Then the equation $12 = 4a_1 + a_0$ has $a_1 = 3$ and $a_0 = 0$ as the solution with the greatest possible value of a_1. The largest 7-10 double is therefore $6 \cdot 49 + 3 \cdot 7 = 315$.

21. If $a \equiv b \pmod{n}$, show that $a^n \equiv b^n \pmod{n^2}$. Is the converse true?

Proof: From $a \equiv b \pmod{n}$ it follows that $a = b + qn$ for some integer q. By the binomial theorem we obtain

$$\begin{aligned} a^n - b^n &= (b + qn)^n - b^n \\ &= \binom{n}{1} b^{n-1} qn + \binom{n}{2} b^{n-2} q^2 n^2 + \cdots + \binom{n}{n} q^n n^n \\ &= n^2 \left(b^{n-1} q + \binom{n}{2} b^{n-2} q^2 + \cdots + \binom{n}{n} q^n n^{n-2} \right), \end{aligned}$$

implying that $a^n \equiv b^n \pmod{n^2}$.

The converse is not true because, for instance, $3^4 \equiv 1^4 \pmod{4^2}$ but $3 \not\equiv 1 \pmod{4}$.

22. Let p be a prime, and let $1 \leq k \leq p - 1$ be an integer. Prove that

$$\binom{p-1}{k} \equiv (-1)^k \pmod{p}.$$

First Proof: We induct on k. The conclusion is clearly true for $k = 1$, since

$$\binom{p-1}{1} = p - 1 \equiv -1 \pmod{p}.$$

Assume that the conclusion is true for $k = i$, where $1 \le i \le p - 2$. It is well known (and easy to check by direct computation) that

$$\binom{p-1}{i} + \binom{p-1}{i-1} = \binom{p}{i}.$$

By Corollary 1.10, we have

$$\binom{p-1}{i} + \binom{p-1}{i-1} \equiv 0 \pmod{p}.$$

By the induction hypothesis, we have

$$\binom{p-1}{i} \equiv -\binom{p-1}{i-1} \equiv -(-1)^{i-1} \equiv (-1)^i \pmod{p},$$

completing our induction.

Second Proof: Because

$$\binom{p-1}{k} = \frac{(p-1)(p-2)\cdots(p-k)}{k!}$$

is an integer and $\gcd(k!, p) = 1$, it suffices to show that

$$(p-1)(p-2)\cdots(p-k) \equiv (-1)^k \cdot k! \pmod{p},$$

which is evident.

23. Let p be a prime. Show that there are infinitely many positive integers n such that p divides $2^n - n$.

Proof: If $p = 2$, p divides $2^n - n$ for every even positive integer n. We assume that p is odd. By Fermat's little theorem, $2^{p-1} \equiv 1 \pmod{p}$. It follows that

$$2^{(p-1)^{2k}} \equiv 1 \equiv (p-1)^{2k} \pmod{p};$$

that is, p divides $2^n - n$ for $n = (p-1)^{2k}$.

24. Let n be an integer greater than three. Prove that $1! + 2! + \cdots + n!$ cannot be a perfect power.

Proof: For $n = 4$, we have $1! + 2! + 3! + 4! = 33$, which is not a perfect power. For $k \geq 5$, $k! \equiv 0 \pmod{10}$. It follows that for $n \geq 5$,

$$1! + 2! + 3! + 4! + \cdots + n! \equiv 3 \pmod{10},$$

so it cannot be a perfect square, or an even power, for this reason.

For odd powers, the following argument settles all cases: one checks the claim for $n < 9$ directly; for $k \geq 9$, $k!$ is a multiple of 27, while $1! + 2! + \cdots + 8!$ is a multiple of 9, but not 27. Hence $1! + 2! + \cdots + n!$ cannot be a cube or higher power.

25. Let k be an odd positive integer. Prove that

$$(1 + 2 + \cdots + n) \mid (1^k + 2^k + \cdots + n^k)$$

for all positive integers n.

Proof: We consider two cases.

In the first case, we assume that n is odd and write $n = 2m + 1$. Then $1 + 2 + \cdots + n = (m + 1)(2m + 1)$. We have

$$\begin{aligned}
&1^k + 2^k + \cdots + n^k \\
&\quad = 1^k + 2^k + \cdots + (2m + 1)^k \\
&\quad = [1^k + (2m + 1)^k] + [2^k + (2m)^k] + \cdots + [m^k + (m + 2)^k] \\
&\qquad + (m + 1)^k.
\end{aligned}$$

Since k is odd, $x + y$ is a factor of $x^k + y^k$. Hence $2m + 2$ divides $i^k + (2m + 2 - i)^k$ for $i = 1, 2, \ldots, m$. Consequently, $m + 1$ divides $1^k + 2^k + \cdots + n^k$. Likewise, we have

$$\begin{aligned}
&1^k + 2^k + \cdots + n^k \\
&\quad = 1^k + 2^k + \cdots + (2m + 1)^k \\
&\quad = [1^k + (2m)^k] + [2^k + (2m - 1)^k] + \cdots + [m^k + (m + 1)^k] \\
&\qquad + (2m + 1)^k.
\end{aligned}$$

Hence $2m+1$ divides $i^k + (2m+1-i)^k$ for $i = 1, 2, \ldots, m$. Consequently, $2m + 1$ divides $1^k + 2^k + \cdots + n^k$. We have shown that each of $m + 1$ and $2m + 1$ divides $1^k + 2^k + \cdots + n^k$. Since $\gcd(m + 1, 2m + 1) = 1$, we conclude that $(m + 1)(2m + 1)$ divides $1^k + 2^k + \cdots + n^k$.

In the second case, we assume that n is even. The proof is similar to that of the first case. We leave it to the reader.

26. Let p be a prime greater than 5. Prove that $p-4$ cannot be the fourth power of an integer.

Proof: Assume that $p-4=q^4$ for some positive integer q. Then $p=q^4+4$ and $q>1$. We obtain

$$\begin{aligned} p &= q^4+4q^2+4-4q^2 = (q^2+2)^2-(2q)^2 \\ &= (q^2-2q+2)(q^2+2q+2), \end{aligned}$$

a product of two integers greater than 1, contradicting the fact that p is a prime. (Note that for $p>5$, $q>1$, and so $(q-1)^2=q^2-2q+1>0$, or $q^2-2q+2>1$.)

27. For a positive integer n, prove that

$$\sigma(1)+\sigma(2)+\cdots+\sigma(n)\le n^2.$$

Proof: The ith summand on the left-hand side is the sum of all the divisors of i. If we write out all these summands on the left-hand side explicitly, each number d, with $1\le d\le n$, appears $\lfloor \frac{n}{d}\rfloor$ times, once for each multiple of d that is less than or equal to n. Hence the left-hand side of the desired inequality is equal to

$$\begin{aligned} &1\cdot\left\lfloor\frac{n}{1}\right\rfloor+2\cdot\left\lfloor\frac{n}{2}\right\rfloor+3\cdot\left\lfloor\frac{n}{3}\right\rfloor+\cdots+n\cdot\left\lfloor\frac{n}{n}\right\rfloor \\ &\qquad\le 1\cdot\frac{n}{1}+2\cdot\frac{n}{2}+3\cdot\frac{n}{3}+\cdots+n\cdot\frac{n}{n}=n^2. \end{aligned}$$

28. [APMO 2004] Determine all finite nonempty sets S of positive integers satisfying

$$\frac{i+j}{\gcd(i,j)}$$

is an element of S for all i and j (not necessarily distinct) in S.

Solution: The answer is $S=\{2\}$.

First of all, taking $i=j$ in the given condition shows that $\frac{i+j}{\gcd(i,j)}=\frac{2i}{i}=2$ is in S. We claim that there is no other element in S. Assume to the contrary that S contains elements other than 2. Let s be the smallest element in S that is not equal to 2.

If s is odd, then $\frac{s+2}{\gcd(s,2)} = s+2$ is another odd element in S. In this way, we will have infinitely many odd numbers in S, contradicting to the fact that S is a finite set.

Hence s must be even, and so $s > 2$. Then $\frac{s+2}{\gcd(s,2)} = \frac{s}{2} + 1$ is in S. For $s > 2$, $2 < \frac{s}{2} + 1 < s$, contradicting the minimality assumption of s.

Note: What if i and j are distinct in the given condition? Kevin Modzelewski showed that the answer is all the sets in the form of $\{a+1, a(a+1)\}$, where a is a positive integer. We leave the proof to the reader.

29. Knowing that 2^{29} is a nine-digit number all of whose digits are distinct, without computing the actual number determine which of the ten digits is missing. Justify your answer.

Solution: Note that $2^3 \equiv -1 \pmod 9$, and hence $2^{29} \equiv (2^3)^9 \cdot 2^2 \equiv -4 \equiv 5 \pmod 9$. The ten-digit number containing all digits 0 through 9 is a multiple of 9, because the sum of its digits has this property. So, in our nine-digit number, 4 is missing. (Indeed, $2^{29} = 536870912$.)

30. Prove that for any integer n great than 1, the number $n^5 + n^4 + 1$ is not prime.

Proof: The given expression factors as

$$\begin{aligned} n^5 + n^4 + 1 &= n^5 + n^4 + n^3 - n^3 - n^2 - n + n^2 + n + 1 \\ &= (n^2 + n + 1)(n^3 - n + 1). \end{aligned}$$

Hence for $n > 1$, it is the product of two integers greater than 1.

One senses the lack of motivation for this factoring. Indeed, with a little bit knowledge of complex numbers, we can present solid algebraic and number-theoretic reasoning for this factoring. We know that $x = 1, \omega, \omega^2$, where $\omega = -\frac{1}{2} + \frac{\sqrt{3}i}{2} = \operatorname{cis} 120^\circ$, are the three roots of the equation $x^3 - 1 = (x-1)(x^2 + x + 1) = 0$. More precisely, ω and ω^2 are the two roots of $x^2 - x + 1 = 0$. Since $\omega^3 = 1$, $\omega^5 + \omega^4 + 1 = \omega^2 + \omega + 1 = 0$, it follows that ω and ω^2 are roots of $x^5 + x^4 + 1 = 0$. We conclude that $n^2 + n + 1$ must be a factor of $n^5 + n^4 + 1$. In the light of this argument, we can replace 4 and 5 in the problem statement by any pair of positive integers congruent to 1 and 2 modulo 3.

31. [Hungary 1995] The product of a few primes is ten times as much as the sum of the primes. What are these (not necessarily distinct) primes?

Solution: Obviously 2 and 5 must be among the primes, and there must be at least one more. Let $p_1 < p_2 < \cdots < p_n$ be the remaining primes. By the given conditions, we deduce that

$$p_1 + p_2 + \cdots + p_n + 7 = p_1 p_2 \cdots p_n. \quad (*)$$

The product of any collection of numbers, each at least 2, must be at least as large as their sum. For two numbers x and y this follows because

$$0 \le (x-1)(y-1) - 1 = xy - x - y.$$

The general result follows by applying this fact repeatedly as

$$x_1 x_2 \cdots x_k \ge x_1 x_2 \cdots x_{k-1} + x_k \ge \cdots \ge x_1 + x_2 + \cdots + x_k.$$

In this problem, we have

$$p_1 + p_2 + \cdots + p_n + 7 = p_1 p_2 \cdots p_n \ge (p_1 + p_2 \cdots + p_{n-1}) p_n.$$

Setting $s = p_1 + \cdots + p_{n-1}$, the last equation can be written as $s + p_n + 7 \ge s p_n$, or

$$(s-1)(p_n - 1) \le 8.$$

We can have $s = 0$ only if there are no primes left, in which case equation $(*)$ becomes $p_n + 7 = p_n$, a contradiction. Hence $s \ge 2$ and so we must have $p_n - 1 \le 8$. This leaves $p_n = 2, 3, 5$ as the only options.

If $p_n = 2$, equation $(*)$ becomes $2n + 7 = 2^n$, which is impossible modulo 2.

If $p_n = 3$, then $p_n - 1 = 2$, and so $s - 1 \le 4$. Then $\{p_1, p_2, \ldots, p_{n-1}\}$ can equal only $\{2\}$, $\{3\}$, or $\{2, 2\}$, $\{2, 3\}$. We check easily that none of these sets satisfies the equation $(*)$.

If $p_n = 5$, then $p_n - 1 = 4$, and so $s - 1 \le 2$, and so the remaining primes must be either a single 2 or a single 3. We check easily that only the latter case gives a solution.

Hence the primes in the collection are $\{2, 3, 5, 5\}$.

32. [Russia 1998] A 10-digit number is said to be *interesting* if its digits are all distinct and it is a multiple of 11111. How many interesting integers are there?

Solution: There are 3456 such integers.

Let $n = \overline{abcdefghij}$ be a 10-digit interesting number. The digits of n must be 0,1, ... ,9, so modulo 9,

$$n \equiv a+b+c+d+e+f+g+h+i+j \equiv 0+1+2+\cdots+9 \equiv 0;$$

that is, 9 divides n. Because $\gcd(9, 11111) = 1$, it follows that $99999 = 9 \cdot 11111$ divides n. Let $x = \overline{abcde}$ and $y = \overline{fghij}$ be two 5-digit numbers. We have $n = 10^5 x + y$. Thus

$$0 \equiv n \equiv 10^5 x + y \equiv x + y \pmod{99999}.$$

But $0 < x+y < 2 \cdot 99999$, so n is interesting if and only if $x+y = 99999$, that is, if $a+f = \cdots = e+j = 9$.

There are $5! = 120$ ways to distribute the pairs $(0,9)$, $(1,8)$, ..., $(4,5)$ among (a, f), (b, g), ..., (e, j), and for each pair we can swap the order of the digits: for example, (b, g) could be $(0, 9)$ or $(9, 0)$. This gives $2^5 = 32$ more choices for a total of $32 \cdot 120$ numbers. However, one-tenth of these numbers have $a = 0$, which is not allowed. So, there are $\frac{9}{10} \cdot 32 \cdot 120 = 3456$ interesting numbers, as claimed.

33. [Russia 1999] Do there exist 19 distinct positive integers that add up to 1999 and have the same sum of digits?

Solution: The answer is negative. Suppose, by way of contradiction, that such integers did exist.

The average of the numbers is $\frac{1999}{19} < 106$, so one number is at most 105 and has digit sum at most 18 (for number 99).

Every number is congruent to its digit sum modulo 9, so all the numbers and their digit sums are congruent modulo 9, say congruent to k. Then $k \equiv 19k \equiv 1999 \equiv 1 \pmod 9$, so the common digit sum is either 1 or 10.

If it is 1, then all the numbers are equal to 1, 10, 100, or 1000, so that some two are equal. This is not allowed. Thus the common digit sum is 10. Note that the twenty smallest numbers with digit sum 10 are

$$19, 28, 37, \ldots, 91, 109, 118, 127, \ldots, 190, 208.$$

The sum of the first nine numbers is

$$(10+20+\cdots+90)+(9+8+\cdots+1) = 450+45 = 495,$$

while the sum of the next nine numbers is

$$(900) + (10 + 20 + \cdots + 80) + (9 + 8 + 7 + \cdots + 1) = 900 + 360 + 45$$
$$= 1305.$$

Hence the first eighteen numbers add up to 1800.

Because $1800 + 190 \neq 1999$, the largest number among the nineteen must be at least 208. Hence the smallest eighteen numbers add up to at least 1800, giving a total sum of at least $2028 > 1999$, a contradiction.

34. [Bulgaria 1995] Find all prime numbers p and q for which pq divides the product $(5^p - 2^p)(5^q - 2^q)$.

Solution: The solutions are $(p, q) = (3, 3), (3, 13)$, or $(13, 3)$. It is easy to check that these are solutions.

Now we show that they are the only solutions. By symmetry, we may assume that $p \le q$. Since $(5^p - 2^p)(5^q - 2^q)$ is odd, we have $q \le p \le 3$.

We observe that if a prime k divides $5^k - 2^k$, then by Fermat's little theorem, we have $3 \equiv 5 - 2 \equiv 5^k - 2^k \pmod{k}$, or $k = 3$.

Assume that $p > 3$. By our observation, we have that p divides $5^q - 2^q$, or $5^q \equiv 2^q \pmod{p}$. By Fermat's little theorem, we have $5^{p-1} \equiv 2^{p-1} \pmod{p}$. By Corollary 1.23,

$$5^{\gcd(p-1,q)} \equiv 2^{\gcd(p-1,q)} \pmod{p}.$$

Because $q \ge p$, $\gcd(p - 1, q) = 1$. The last congruence relation now reads $5 \equiv 2 \pmod{p}$, implying that $p = 3$, a contradiction.

Hence $p = 3$. If $q > 3$, by our observation, q must divide $5^p - 2^p = 5^3 - 2^3 = 9 \cdot 13$, and so $q = 13$, leading to the solution $(p, q) = (3, 13)$.

35. Prove that there are infinitely many numbers not containing the digit 0 that are divisible by the sum of their digits.

Proof: For a positive integer n, let

$$a_n = \underbrace{11\ldots1}_{3^n}.$$

It suffices to show that for all positive integers n, a_n is divisible by the sum of its digits; that is, a_n is divisible by 3^n.

We induct on n. For $n = 1$, it is clear that $a_n = 111$, which is divisible by 3. Assume that a_n is divisible by 3^n, for some positive integer $n = k$. We consider a_{k+1}. Note that

$$\begin{aligned} a_{k+1} &= \underbrace{11\ldots1}_{3^{k+1}} = \underbrace{11\ldots1}_{3^k\cdot 3} = \underbrace{11\ldots1}_{3^k}\underbrace{11\ldots1}_{3^k}\underbrace{11\ldots1}_{3^k} \\ &= \underbrace{11\ldots1}_{3^k}\left(10^{2\cdot 3^k} + 10^{3^k} + 1\right) \\ &= a_k \cdot 1\underbrace{0\ldots0}_{3^k-1}1\underbrace{0\ldots0}_{3^k-1}1. \end{aligned}$$

Because 3 divides $1\underbrace{0\ldots0}_{3^k-1}1\underbrace{0\ldots0}_{3^k-1}1$ and 3^k divides a_k, it follows that 3^{k+1} divides a_{k+1}. This completes our induction.

36. Prove that any number consisting of 2^n identical digits has at least n distinct prime factors.

Proof: Such a number N can be written as

$$N = k \cdot \frac{10^{2^n} - 1}{10 - 1} = k(10 + 1)(10^2 + 1)\cdots(10^{2^{n-1}} + 1).$$

The desired conclusion follows from the fact that the n factors $10^{2^h} + 1$, $h = 0, 1, \ldots, n - 1$, are pairwise relatively prime. Indeed, for $h_1 > h_2$,

$$\begin{aligned} 10^{2^{h_2}} + 1 \mid 10^{2^{h_1}} - 1 \\ &= 9 \cdot (10 + 1)(10^2 + 1)\cdots(10^{2^{h_2}} + 1)\cdots(10^{2^{h_1-1}} + 1), \end{aligned}$$

so

$$\begin{aligned} \gcd(10^{2^{h_2}} + 1, 10^{2^{h_1}} + 1) &= \gcd(10^{2^{h_1}} - 1, 10^{2^{h_1}} + 1) \\ &= \gcd(2, 10^{2^{h_1}} + 1) = 1. \end{aligned}$$

Note: There is another way to see that $\gcd(10^{2^{h_1}} + 1, 10^{2^{h_2}} + 1) = 1$. If p divides $10^{2^{h_2}} + 1$, then p must be odd. Since $10^{2^{h_2}} \equiv -1 \pmod{p}$, it follows that

$$10^{2^{h_1}} \equiv \left(10^{2^{h_2}}\right)^{2^{h_1-h_2}} \equiv (-1)^{2^{h_1-h_2}} \equiv 1 \pmod{p},$$

implying that p divides $10^{2^{h_1}} - 1$. Since p is odd, p does not divide $10^{2^{h_1}} + 1$.

37. Let a and b be two relatively prime positive integers, and consider the arithmetic progression $a, a+b, a+2b, a+3b, \ldots$.

    (1) [G. Polya] Prove that there are infinitely many terms in the arithmetic progression that have the same prime divisors.

    (2) Prove that there are infinitely many pairwise relatively prime terms in the arithmetic progression.

First Proof: In this approach, we apply properties of linear congruences.

(1) Since $\gcd(a, b) = 1$, a has an inverse modulo b. Let x be a positive integer such that $ax \equiv 1 \pmod{b}$. For every positive integer n, let $s_n = (a+b)(ax)^n$. Then $s_n \equiv a \pmod{b}$; that is, s_n is a term in the arithmetic progression. It is clear that these terms have the same prime divisors, namely, the divisors of a, x, and $a+b$.

(2) We construct these terms inductively with the additional condition that these terms are relatively prime to a. Let $t_1 = a+b$. Then $\gcd(t_1, t_2) = 1$ and $\gcd(t_1, a) = 1$. Assume that terms $t_1, \ldots, t_k$ have been chosen such that $\gcd(t_i, t_j) = 1$ and $\gcd(a, t_i) = 1$ for $1 \le i < j \le k$. Set

$$t_{k+1} = t_1 \cdots t_k b + a.$$

Clearly, t_{k+1} is a term in the arithmetic progression. Because $t_1, \ldots,$ t_k are distinct integers greater than 1, it is not difficult to see that $t_{k+1} > t_i$, $1 \le i \le k$. It is also not difficult to see that $\gcd(t_{k+1}, a) = 1$ by the induction hypothesis and the given condition $\gcd(a, b) = 1$. It remains to show that $\gcd(t_{k+1}, t_i) = 1$ for $1 \le i \le k$, which follows from

$$\gcd(t_{k+1}, t_i) = \gcd(t_1 t_2 \cdots t_k b + a, t_i) = \gcd(a, t_i) = 1,$$

again by the induction hypothesis. Our induction is thus complete.

Second Proof: (By Sherry Gong) In this approach, we apply Euler's theorem.

(1) The terms $x_n = (a+b)^{n\varphi(b)+1}$ satisfy the conditions of the problem. We note that these terms share the same prime divisors, namely, the divisors of $a+b$. It remains to show that x_n appears in the arithmetic progression for each large integer n. By Euler's theorem, we have

$$x_n \equiv a^{n\varphi(b)+1} \equiv a^{n\varphi(b)} \cdot a \equiv a \pmod{b}.$$

Hence $x_n = a + kb$. For large n, x_n must appears in the given arithmetic progression.

(2) Let $y_1 = a$ and $y_2 = a + b$. Clearly, $\gcd(y_1, y_2) = 1$. Assume that we have pairwise relatively prime terms $y_1 < y_2 < \cdots < y_k$ in the sequence. We set

$$y_{k+1} = y_1 y_2 \cdots y_k a^{z_{k+1}\varphi(b)-k+1} + b,$$

where z_{k+1} is some large integer such that $y_{k+1} > y_k$. We claim that y_{k+1} is a term in the arithmetic progression that is relatively prime to each of $y_1, y_2, \ldots, y_k$. In this way, we can construct one new term at a time inductively to produce a subsequence of the arithmetic progression satisfying the conditions of the problem.

Now we prove our claim. We note that

$$y_{k+1} \equiv a^k a^{z_{k+1}\varphi(b)-k+1} \equiv a \pmod{b},$$

implying that y_{k+1} is a term in the arithmetic progression. For each $1 \leq i \leq k$, we also have

$$\gcd(y_{k+1}, y_i) = \gcd(b, y_i) = \gcd(b, a) = 1,$$

by noting that y_i is a term in the arithmetic progression. Our proof is thus complete.

Note: We can slightly modify our proof so that the conclusions hold for all relatively prime integers a and b. In part (1), $\gcd(a, b)$ does not even need to be 1. Because by factoring out $\gcd(a, b)$ from each term in the progression we reduce to the current part (1).

38. Let n be a positive integer.

(1) Evaluate $\gcd(n! + 1, (n + 1)! + 1)$.

(2) Let a and b be positive integers. Prove that

$$\gcd(n^a - 1, n^b - 1) = n^{\gcd(a,b)} - 1.$$

(3) Let a and b be positive integers. Prove that $\gcd(n^a + 1, n^b + 1)$ divides $n^{\gcd(a,b)} + 1$.

(4) Let m be a positive integer with $\gcd(m, n) = 1$. Express

$$\gcd(5^m + 7^m, 5^n + 7^n)$$

in terms of m and n.

Proof: We apply the Euclidean algorithm and Corollary 1.23 to this problem.

(1) By the Euclidean algorithm, we have

$$\begin{aligned}\gcd(n!+1, (n+1)!+1)& \\ &= \gcd(n!+1, (n+1)!+1-(n+1)(n!+1)) \\ &= \gcd(n!+1, n) = 1.\end{aligned}$$

(2) Without loss of generality, we assume that $a \geq b$. Then

$$\begin{aligned}\gcd(n^a-1, n^b-1) &= \gcd(n^a-1-n^{a-b}(n^b-1), n^b-1) \\ &= \gcd(n^{a-b}-1, n^b-1).\end{aligned}$$

Recall the process of finding $\gcd(a, b) = \gcd(a-b, b)$. We see that the process of computing $\gcd(n^a-1, n^b-1)$ is the same as the process of computing $\gcd(a, b)$ as the exponents, from which the conclusion follows.

Alternatively, we can also approach the problem the following way. Since $\gcd(a, b)$ divides both a and b, the polynomial $x^{\gcd(a,b)}-1$ divides both x^a-1 and x^b-1. Hence $n^{\gcd(a,b)}-1$ divides both n^a-1 and n^b-1, implying that

$$n^{\gcd(a,b)}-1 \mid \gcd(n^a-1, n^b-1).$$

On the other hand, assume that m divides both n^a-1 and n^b-1; that is, $n^a \equiv 1 \equiv 1^a \pmod{m}$ and $n^b \equiv 1 \equiv 1^b \pmod{m}$ (clearly, m and n are relatively prime to each other). By Corollary 1.23, we have $n^{\gcd(a,b)} \equiv 1 \pmod{m}$; that is, m divides $n^{\gcd(a,b)}-1$. It follows that

$$\gcd(n^a-1, n^b-1) \mid n^{\gcd(a,b)}-1.$$

we conclude that $n^{\gcd(a,b)}-1 = \gcd(n^a-1, n^b-1)$.

(3) Assume that m divides both 2^a+1 and 2^b+1. Note that m is odd. It suffices to show that m divides $2^{\gcd(a,b)}+1$.

Since $2^a \equiv 2^b \equiv -1 \pmod{m}$, we have

$$2^{2a} \equiv 1 \pmod{m} \quad \text{and} \quad 2^{2b} \equiv 1 \pmod{m}.$$

By Corollary 1.23, it follows that $2^{\gcd(2a,2b)} \equiv 1 \pmod{m}$; that is, m divides $2^{\gcd(2a,ab)}-1 = 2^{2\gcd(a,b)}-1$, or

$$m \mid (2^{\gcd(a,b)}-1)(2^{\gcd(a,b)}+1).$$

If m divides $2^{\gcd(a,b)}+1$, we are done. Assume that m does not divide $2^{\gcd(a,b)}+1$. Since

$$\gcd(2^{gcd(a,b)}-1, 2^{gcd(a,b)}+1) = \gcd(2, 2^{gcd(a,b)}-1) = 1,$$

m must divide $2^{\gcd(a,b)-1}$, which divides 2^a-1 (as we showed in the proof of (3)). But m divides 2^a+1 by our original assumption. Thus m divides $\gcd(2^a+1, 2^a-1) = 2$. Since m is odd, $m = 1$, contradicting the assumption that m does not divide $2^{\gcd(a,b)}+1$. Thus, m must divide $2^{\gcd(a,b)}+1$, completing our proof.

(4) Let $s_n = 5^n + 7^n$. If $n \geq 2m$, note that

$$s_n = s_m s_{n-m} - 5^m 7^m s_{n-2m},$$

so $\gcd(s_m, s_n) = \gcd(s_m, s_{n-2m})$.
Similarly, if $m < n < 2m$, we have

$$s_n = s_m s_{n-m} - 5^{n-m} 7^{n-m} s_{2m-n},$$

so $\gcd(s_m, s_n) = \gcd(s_m, s_{2m-n})$.
Thus by the Euclidean algorithm, we conclude that if $m+n$ is even, then $\gcd(s_m, s_n) = \gcd(s_1, s_1) = 12$, and if $m+n$ is odd, then $\gcd(s_m, s_n) = \gcd(s_0, s_1) = 2$.

Note: The interested reader might want to generalize part (3), considering the relation between $\gcd(n^a+1, n^b+1)$ and $n^{\gcd(a,b)}+1$.

39. Bases? What bases?

(1) Determine whether it is possible to find a cube and a plane such that the distances from the vertices of the cube to the plane are $0, 1, 2, \ldots, 7$.

(2) [AIME 1986] The increasing sequence $1, 3, 4, 9, 10, 12, 13, \ldots$ consists of all those positive integers that are powers of 3 or sums of distinct powers of 3. Find the 100th term of this sequence (where 1 is the 1st term, 3 is the 2nd term, and so on).

Solution: In this problem, we apply base 2 (binary representation) and base 4.

(1) The answer is positive.
We consider a unit cube $\mathcal{S}$ with vertices $(0, 0, 0)$, $(0, 0, 1)$, $(0, 1, 0)$, $(0, 1, 1)$, $(1, 0, 0)$, $(1, 0, 1)$, $(1, 1, 0)$, and $(1, 1, 1)$. We note that these

coordinates match the binary representation of 0, 1, 2, 3, 4, 5, 6, 7. This motivates us to consider the plane $x+2y+4z = 0$. The distances from the vertices of $\mathcal{S}$ to the plane are

$$0, \quad \frac{1}{\sqrt{21}}, \quad \frac{2}{\sqrt{21}}, \quad \frac{3}{\sqrt{21}}, \quad \frac{4}{\sqrt{21}}, \quad \frac{5}{\sqrt{21}}, \quad \frac{6}{\sqrt{21}}, \quad \frac{7}{\sqrt{21}}.$$

By a simple scaling, we can find a cube satisfying the conditions of the problem. Indeed, we dilate S via the origin with a ratio of $\sqrt{21}$ to obtain cube $\mathcal{T}$. Point (a, b, c) maps to point $(\sqrt{21}a, \sqrt{21}b, \sqrt{21}c)$. Then cube $\mathcal{T}$ and plane $x + 2y + 4z = 0$ satisfy the conditions of the problem.

(2) Note that a positive integer is a term of this sequence if and only if its base-3 representation consists only of 0's and 1's. Therefore, we can set up a one-to-one correspondence between the positive integers and the terms of this sequence by representing both with binary digits (0's and 1's), first in base 2 and then in base 3:

$$\begin{aligned} 1 &= 1_{(2)} && \Longleftrightarrow 1_{(3)} = 1, \\ 2 &= 10_{(2)} && \Longleftrightarrow 10_{(3)} = 3, \\ 3 &= 11_{(2)} && \Longleftrightarrow 11_{(3)} = 4, \\ 4 &= 100_{(2)} && \Longleftrightarrow 100_{(3)} = 9, \\ 5 &= 101_{(2)} && \Longleftrightarrow 101_{(3)} = 10, \\ &\vdots \end{aligned}$$

This is a correspondence between the two sequences in the order given, that is, the kth positive integer is made to correspond to the kth sum (in increasing order) of distinct powers of 3. This is because when the binary numbers are written in increasing order, they are still in increasing order when interpreted in any other base.

Therefore, to find the 100th term of the sequence, we need only look at the 100th line of the above correspondence:

$$100 = 1100100_{(2)} \Longleftrightarrow 1100100_{(3)} = 981.$$

Note: The key facts in the solution of part (3) are the following: (a) for integers a and b in X, no carries appear (among the digits) in the addition $a + 2b$; (b) each digit in base 4 (namely, 0, 1, 2, 3) can be uniquely written in the form of $a + 2b$, where a and b are equal to either 0 or 1. In other words, we can uniquely write a base-4 digit in base 2:

base 2	$00_{(2)}$	$01_{(2)}$	$10_{(2)}$	$11_{(2)}$
base 4	$0_{(4)}$	$1_{(4)}$	$2_{(4)}$	$3_{(4)}$

Based on this, for each nonnegative integer $n = \overline{n_k n_{k-1} \dots a_0}_{(4)}$, we can find the solution $a + 2b = n$ in this way: write each digit n_i in base 2 according the above table to obtain the binary representation of n. The digits in odd positions from the left form the base-4 representation of a, and the digits in the even positions from the left form the base-4 representation of b. For example,

$$\begin{aligned} 123210_{(4)} &= \mathbf{0}1, \mathbf{1}0, \mathbf{1}1, \mathbf{1}0, \mathbf{0}1, \mathbf{0}0_{(2)} \\ &= 101010_{(4)} + 2 \cdot 011100_{(4)} = 101010_{(4)} + 2 \cdot 11100_{(4)}. \end{aligned}$$

40. Fractions in modular arithmetic.

(1) [ARML 2002] Let a be the integer such that

$$1 + \frac{1}{2} + \frac{1}{3} + \cdots + \frac{1}{23} = \frac{a}{23!}.$$

Compute the remainder when a is divided by 13.

(2) Let $p > 3$ be a prime, and let m and n be relatively prime integers such that

$$\frac{m}{n} = \frac{1}{1^2} + \frac{1}{2^2} + \cdots + \frac{1}{(p-1)^2}.$$

Prove that m is divisible by p.

(3) [Wolstenholme's Theorem] Let $p > 3$ be a prime. Prove that

$$p^2 \mid (p-1)! \left(1 + \frac{1}{2} + \cdots + \frac{1}{p-1}\right).$$

Solution:

(1) Note that

$$a = 23! + \frac{23!}{2} + \cdots + \frac{23!}{23}.$$

Besides $\frac{23!}{13}$, each summand on the right-hand side is an integer divisible by 13. Hence, by Wilson's theorem, we have

$$\begin{aligned} a \equiv \frac{23!}{13} &\equiv 12! \cdot 14 \cdot 15 \cdots 23 \\ &\equiv 12!10! \equiv \frac{(12!)^2}{11 \cdot 12} \equiv \frac{1}{2} \equiv 7 \pmod{13}. \end{aligned}$$

(2) Note that

$$((p-1)!)^2\frac{m}{n} = ((p-1)!)^2\left(\frac{1}{1^2}+\frac{1}{2^2}+\cdots+\frac{1}{(p-1)^2}\right)$$

is an integer. Note also that

$$\left\{\frac{1}{1},\frac{1}{2},\ldots,\frac{1}{p-1}\right\}$$

is a reduced complete set of residue classes modulo p. By Proposition 1.18 (h) and Wilson's theorem, we have

$$\begin{aligned}((p-1)!)^2&\left(\frac{1}{1^2}+\frac{1}{2^2}+\cdots+\frac{1}{(p-1)^2}\right)\\ &\equiv (-1)^2[1^2+2^2+\cdots+(p-1)^2]\\ &\equiv \frac{(p-1)p(2p-3)}{6} \equiv 0 \pmod{p},\end{aligned}$$

since $p \geq 5$ and so $\gcd(6, p) = 1$. Hence p divides the integer $\frac{((p-1)!)^2 m}{n}$. Since $\gcd((p-1)!, p) = 1$, we must have $p \mid m$, as desired.

(3) Set

$$S = (p-1)!\left(1+\frac{1}{2}+\cdots+\frac{1}{p-1}\right).$$

Then

$$2S = (p-1)!\sum_{i=1}^{p-1}\left[\frac{1}{i}+\frac{1}{p-i}\right] = (p-1)!\sum_{i=1}^{p-1}\frac{p}{i(p-i)} = p\cdot T,$$

where

$$T = (p-1)!\sum_{i=1}^{p-1}\frac{1}{i(p-i)}.$$

Since $2S$ is an integer and p is relatively prime to the numerators of the summands in T, T must itself be an integer. Since $p > 3$, $\gcd(p, 2) = 1$ and p must divide S. It suffices to show that p also divides T. By (2), we have

$$T \equiv (p-1)!\sum_{i=1}^{p-1} -\frac{1}{i^2} \equiv (p-1)!\frac{m}{n} \equiv 0 \pmod{p},$$

since $p \mid m$ and $\gcd(m, n) = 1$.

41. Find all pairs (x, y) of positive integers such that $x^2 + 3y$ and $y^2 + 3x$ are simultaneously perfect squares.

Solution: The answers are $(x, y) = (1, 1), (11, 16)$, or $(16, 11)$. It is easy to check that they are solutions. We show that they are the only answers.

The inequalities

$$x^2 + 3y \geq (x + 2)^2 \quad \text{and} \quad y^2 + 3x \geq (y + 2)^2$$

cannot hold simultaneously because summing them up yields $0 \geq x+y+8$, which is false. Hence at least one of $x^2 + 3y < (x + 2)^2$ and $y^2 + 3x < (y+2)^2$ is true. Without loss of generality assume that $x^2+3y < (x+2)^2$. From $x^2 < x^2 + 3y < (x + 2)^2$ we derive $x^2 + 3y = (x + 1)^2$; hence $3y = 2x + 1$. Then $x = 3k + 1$ and $y = 2k + 1$ for some nonnegative integer k. Consequently, we have $y^2 + 3x = 4k^2 + 13k + 4$. If $k > 5$, then

$$(2k + 3)^2 < 4k^2 + 13k + 4 < (2k + 4)^2,$$

and so $y^2 + 3x$ cannot be a square. It is not difficult to check that for $k \in \{1, 2, 3, 4\}$, $y^2 + 3x$ is not a perfect square and that for $k = 0$, $y^2 + 3x = 4 = 2^2$ and for $k = 5$, $y^2 + 3x = 13^2$. For these two values of k, $x^2 + 3y$ is equal to 2^2 or 17^2, leading to solutions $(x, y) = (1, 1)$ and $(x, y) = (16, 11)$.

42. First digit? Not the last digit? Are your sure?

(1) [AMC12B 2004] Given that 2^{2004} is a 604-digit number with leading digit 1, determine the number of elements in the set

$$\{2^0,\ 2^1,\ 2^2,\ \ldots,\ 2^{2003}\}$$

with leading digit 4.

(2) Let k be a positive integer and let $n = n(k)$ be a positive integer such that in decimal representation 2^n and 5^n begin with the same k digits. What are these digits?

Solution: We present two solutions for part (1).

(1) • *First approach.* The smallest power of 2 with a given number of digits has a first digit (most left digit) of 1, and there are elements of S with n digits for each integer $n \leq 603$, so there are 603 elements of S whose first digit is 1. Furthermore, if the first digit of 2^k is 1, then the first digit of 2^{k+1} is either 2 or 3, and the

first digit of 2^{k+2} is either 4, 5, 6, or 7. Therefore there are 603 elements of S whose first digit is 2 or 3, 603 elements whose first digits is 4, 5, 6, or 7, and $2004 - 3(603) = 195$ elements whose first digit is 8 or 9. Finally, note that the first digit of 2^k is 8 or 9 if and only if the first digit of 2^{k-1} is 4, so there are 195 elements of S whose first digit is 4.

- *Second approach.* We partition the set S into the following blocks:

$$\{2^0, 2^1, 2^2, 2^3; \quad 2^4, 2^5, 2^6; \quad , 2^7, \ldots, 2^{2003}\},$$

where the leading term in each block has first digit 1. Because 2^{2004} has first digit 1, S has been partitioned into complete blocks. As we showed in the first approach, there are exactly 603 elements in S whose first digit is 1. Hence there are 603 blocks in S. Note that a block can have either 3 or 4 elements. If a block has 3 elements 2^k, 2^{k+1}, and 2^{k+2}, the their first digits are 1, 2 or 3, 5 or 6 or 7; if a block has 4 elements 2^k, 2^{k+1}, 2^{k+2}, and 2^{k+3}, then their first digits are 1, 2, 4, 8, or 9. Thus the number of elements in S having first digit 4 is equal to the number of 4-element blocks. Suppose that there are x 3-element blocks and y 4-element blocks. We have $3x + 4y = 2004$ (since there is a total of 2004 elements in S) and $x + y = 603$ (since there are 603 complete blocks). Solving the equations gives $x = 408$ and $y = 195$.

(2) Let s and t be unique positive integers such that $10^s < 2^n < 10^{s+1}$ and $10^t < 5^n < 10^{t+1}$. Set $a = \frac{2^n}{10^s}$ and $b = \frac{5^n}{10^t}$. Clearly, $1 < a < 10$, $1 < b < 10$, and $ab = 10^{n-s-t}$. Hence ab is a power of 10 and since $1 < ab < 10^2$, the only possibility is $ab = 10$. We obtain

$$\min(a, b) < \sqrt{ab} = \sqrt{10} < \max(a, b),$$

implying that the first common k digits are the first k digits of $\sqrt{10}$. (For $k = 1$, $2^5 = 32$ and $5^5 = 3125$ have the same leading digit, the first digit of $\sqrt{10} = 3.1\ldots$.)

43. What are those missing digits?

(1) Determine the respective last digit (unit digit) of the numbers

$$3^{1001} 7^{1002} 13^{1003} \quad \text{and} \quad \underbrace{7^{7^{7^{\cdot^{\cdot^{\cdot^{7}}}}}}}_{1001\ 7\text{'s}}.$$

(2) [Canada 2003] Determine the last three digits of the number

$$2003^{2002^{2001}}.$$

(3) The binomial coefficient $\binom{99}{19}$ is a 21-digit number:

$$107{,}196{,}674{,}080{,}761{,}936{,}xyz.$$

Find the three-digit number xyz.

(4) Find the smallest positive integer whose cube ends in 888.

First Solution: The key values in this problem are $\varphi(10) = 4$ and $\varphi(1000) = 400$. We repeatedly apply Euler's theorem.

(1) The answers are 9 and 3, respectively.
Note that

$$\begin{aligned} 3^{1001}7^{1002}13^{1003} &\equiv 3^{1000}91^{1002} \cdot 3 \cdot 13 \\ &\equiv 81^{250}91^{1002} \cdot 39 \equiv 9 \pmod{10}. \end{aligned}$$

Since $7^4 \equiv 1 \pmod{10}$, we obtain

$$\underbrace{7^{7^{7^{\cdot^{\cdot^{\cdot^{7}}}}}}}_{1000\ 7\text{'s}} \equiv 3 \pmod 4$$

by noting that $7^{2k} \equiv 1 \pmod 4$ and $7^{2k+1} \equiv 3 \pmod 4$. Hence

$$\underbrace{7^{7^{7^{\cdot^{\cdot^{\cdot^{7}}}}}}}_{1001\ 7\text{'s}} \equiv 7^3 \equiv 3 \pmod{10}.$$

(2) The answer is 241.
Since $\varphi(1000) = 400$ and

$$2003^{2002^{2001}} \equiv 3^{2002^{2001}} \pmod{1000},$$

we need to compute 2002^{2001} modulo 400, or 2^{2001} modulo 400. Since $400 = 16 \cdot 25$ and 16 clearly divides 2^{2001}, $2^{2001} \equiv 16k \pmod{400}$ for some positive integer k. By Corollary 1.21, we deduce $2^{1997} \equiv k \pmod{25}$. Since $\varphi(25) = 20$,

$$k \equiv 2^{1997} \equiv \frac{2^{2000}}{2^3} \equiv \frac{1}{8} \equiv 22 \pmod{25},$$

or $k = 22$. It follows that $2002^{2001} \equiv 2^{2001} \equiv 16k \equiv 352 \pmod{400}$, and so

$$2003^{2002^{2001}} \equiv 3^{2002^{2001}} \equiv 3^{352} \equiv 9^{176} \equiv (10-1)^{176} \pmod{1000}.$$

By the binomial theorem, we have

$$\begin{aligned}(10-1)^{176} &\equiv \binom{176}{2} \cdot 10^2 - \binom{176}{1} \cdot 10 + 1^{176} \\ &\equiv 0 - 760 + 1 \equiv 241 \pmod{1000}.\end{aligned}$$

(3) The answer is 594.

We have

$$\binom{99}{19} = \frac{99!}{19!80!} = \frac{99 \cdot 98 \cdots 81}{19!}.$$

Since $1000 = 8 \cdot 125$, we need to compute $\binom{99}{19}$ modulo 8 and 125, respectively. Instead, we first compute $\binom{99}{19}$ modulo 4 and 25, since 99 is very close to 100. (That is, we compute the y and z first.)

We note that

$$\begin{aligned}&\frac{99 \cdot 98 \cdots 81}{19!} \\ &= \frac{99 \cdot 98 \cdots 96 \cdot 95 \cdot 94 \cdots 91 \cdot 90 \cdot 89 \cdots 86 \cdot 85 \cdot 84 \cdots 81}{4! \cdot 5 \cdot 6 \cdots 9 \cdot 10 \cdot 11 \cdots 14 \cdot 15 \cdot 16 \cdots 19} \\ &= \frac{19 \cdot 18 \cdot 17 \cdot 99 \cdots 96 \cdot 94 \cdots 91 \cdot 89 \cdots 86 \cdot 84 \cdots 81}{3!4! \cdot 6 \cdots 9 \cdot 11 \cdots 14 \cdot 16 \cdots 19}.\end{aligned}$$

Consequently,

$$\frac{99 \cdot 98 \cdots 81}{19!} \equiv \frac{19 \cdot 18 \cdot 17}{3!} \equiv 19 \pmod{25}.$$

In a similar fashion, we can compute $\binom{99}{19}$ modulo 4. Note that

$$\sum_{n=1}^{\infty} \left\lfloor \frac{99}{2^n} \right\rfloor - \sum_{n=1}^{\infty} \left\lfloor \frac{19}{2^n} \right\rfloor - \sum_{n=1}^{\infty} \left\lfloor \frac{80}{2^n} \right\rfloor = 95 - 16 - 78 = 1,$$

from which it follows that $\binom{99}{19} \equiv 2 \pmod{4}$.

Combining the above, we conclude that $\binom{99}{19} \equiv 94 \pmod{100}$; hence that $y = 9$ and $z = 4$.

We also compute

$$e_3\left(\binom{99}{19}\right) = \sum_{n=1}^{\infty}\left\lfloor \frac{99}{3^n} \right\rfloor - \sum_{n=1}^{\infty}\left\lfloor \frac{19}{3^n} \right\rfloor - \sum_{n=1}^{\infty}\left\lfloor \frac{80}{3^n} \right\rfloor$$
$$= 48 - 8 - 36 = 4,$$

from which it follows that $\binom{99}{19} \equiv 0 \pmod{9}$. Hence, modulo 9, we have

$$1+0+7+1+9+6+6+7+4+0+8$$
$$+0+7+6+1+9+3+6+x+9+4 \equiv 0,$$

or $x \equiv 5 \pmod{9}$. Because x is a digit in decimal representation, $x = 5$.

(4) The answer is 192.

If the cube of an integer ends in 8, then the integer itself must end in 2; that is, it must be of the form $10k + 2$. Therefore,

$$n^3 = (10k+2)^3 = 1000k^3 + 600k^2 + 120k + 8,$$

where the penultimate term, $120k$, determines the penultimate digit (tenth digit) of n^3, which must also be 8. In other words,

$$88 \equiv n^3 \equiv 120k + 8 \pmod{100},$$

or $80 \equiv 120k \pmod{100}$. In view of this, by Corollary 1.21, $8 \equiv 12k \pmod{10}$, or $4 \equiv 6k \pmod{5}$. Consequently, $4 \equiv k \pmod{5}$, or $k = 5m + 4$. Thus, modulo 1000, we have

$$888 \equiv n^3 \equiv 600(5m+4)^2 + 120(5m+4) + 8$$
$$\equiv 9600 + 600m + 488,$$

or $800 \equiv 600m \pmod{1000}$. Consequently, by Corollary 1.21 again, we have $8 \equiv 6m \pmod{10}$, or $4 \equiv 3m \pmod{5}$. This leads to $m \equiv 3 \pmod{5}$.

The smallest m that will ensure this is $m = 3$, implying that $k = 5 \cdot 3 + 4 = 19$, and $n = 10 \cdot 19 + 2 = 192$. (Indeed, $192^3 = 7077888$.)

Second Solution: We present another approach for part (3). Similar to what we have shown in the first solution, it is not difficult to prove that $11 \mid \binom{99}{19}$ and $7 \mid \binom{99}{19}$. Applying Proposition 1.44 (b), (c), and (d) leads to

$$x + y + z \equiv 0 \pmod{9},$$
$$x - y + z \equiv 0 \pmod{11},$$
$$\overline{xyz} + 1 \equiv 0 \pmod{7},$$

or

$$x + y + z \equiv 0 \pmod{9},$$
$$x - y + z \equiv 0 \pmod{11},$$
$$2x + 3y + z + 1 \equiv 0 \pmod{7}.$$

Because x, y, and z are digits, the first equation leads to $x + y + z = 9$, or 18, or 27 (with $x = y = z = 9$); and the second equation leads to $x - y + z = 0$ or 11. It is not difficult to see that $(x + y + z, x - y + z) = (18, 0)$, or $(x+z, y) = (9, 9)$. Substituting this into the third equation gives $0 \equiv x + 3y + (x + z) + 1 \equiv x + 2 \pmod{7}$, implying that $x = 5$, and so $z = 4$ and $\overline{xyz} = 594$.

Note: A common mistake in solving part (3) goes as follows:

$$\binom{99}{19} = \frac{99 \cdot 98 \cdots 81}{19!} \equiv \frac{19 \cdot 18 \cdots 1}{19!} \equiv 1 \pmod{8}.$$

Because 19! is not relatively prime to 8, we cannot operate division in this congruence. (Please see the discussion leading to Corollary 1.21.)

44. Let $p \geq 3$ be a prime, and let

$$\{a_1, a_2, \ldots, a_{p-1}\} \quad \text{and} \quad \{b_1, b_2, \ldots, b_{p-1}\}$$

be two sets of complete residue classes modulo p. Prove that

$$\{a_1b_1, a_2b_2, \ldots, a_{p-1}b_{p-1}\}$$

is not a complete set of residue classes modulo p.

Proof: By Wilson's theorem,

$$a_1a_2 \cdots a_{p-1} \equiv b_1b_2 \cdots b_{p-1} \equiv (p-1)! \equiv -1 \pmod{p}.$$

It follows that

$$(a_1b_1)(a_2b_2) \cdots (a_{p-1}b_{p-1})$$
$$\equiv a_1a_2 \cdots a_{p-1}b_1b_2 \cdots b_{p-1} \equiv (-1)^2 \equiv 1 \pmod{p-1},$$

and so

$$\{a_1b_1, a_2b_2, \ldots, a_{p-1}b_{p-1}\}$$

is not a complete set of residue classes modulo p, by Wilson's theorem again.

45. Let $p \geq 3$ be a prime. Determine whether there exists a permutation

$$(a_1, a_2, \ldots, a_{p-1})$$

of $(1, 2, \ldots, p-1)$ such that the sequence $\{ia_i\}_{i=1}^{p-1}$ contains $p-2$ distinct congruence classes modulo p.

Solution: The answer is positive.

For each $1 \leq i \leq p-2$, since $\gcd(i, p) = 1$, i is invertible modulo p, and so $ix \equiv i+1 \pmod{p}$ has a unique solution (modulo p). Let a_i be the unique integer with $1 \leq a_i \leq p-1$ such that $ia_i \equiv i+1 \pmod{p}$. It remains to show that for $1 \leq i < j \leq p-2$, $a_i \neq a_j$. Assume to the contrary that $a_i = a_j = a$ for $1 \leq i < j \leq p-2$. Because

$$ia_i \equiv i+1 \pmod{p} \quad \text{and} \quad ja_j \equiv j+1 \pmod{p},$$

it follows that

$$0 \equiv a(j-i) \equiv ja_j - ia_i \equiv j - i \pmod{p},$$

which is impossible since $0 < j - i < p - 2$.

Note: By Problem 43, $\{a_1, 2a_2, 3a_3, \ldots, (p-1)a_{p-1}\}$ is not a complete set of residue classes. By Problem 44, we conclude that the maximum number of distinct congruence classes in the sequence $\{a_1, 2a_2, 3a_3, \ldots, (p-1)a_{p-1}\}$ is $p-2$.

46. [Paul Erdös] Prove that any positive integer less than $n!$ can be represented as a sum of no more than n positive integer divisors of $n!$.

First Proof: for each $k - 1, 2, \ldots, n$, let $a_k = \frac{n!}{k!}$. Suppose we have some number m with $a_k \leq m < a_{k-1}$, where $2 \leq k \leq n$. Then consider the number $d = a_k \lfloor \frac{m}{a_k} \rfloor$. We have $0 \leq m - d < a_k$; furthermore, because $s = \lfloor \frac{m}{a_k} \rfloor < \frac{a_{k-1}}{a_k} = k$, we know that

$$\frac{n!}{d} = \frac{a_k k!}{a_k s} = \frac{k!}{s}$$

is an integer. Thus from m we can subtract d, a factor of $n!$, to obtain a number less than a_k.

Then if we start with any positive integer $m < n! = a_1$, then by subtracting at most one factor of $n!$ from m we can obtain an integer less than a_2; by subtracting at most one more factor of $n!$ we can obtain an integer less than a_3; and so on, so that we can represent m as the sum of at most $n-1$ positive integer divisors of $n!$.

Second Proof: We proceed by induction. For $n = 3$ the claim is true. Assume that the hypothesis holds for $n - 1$. Let $1 < k < n!$ and let k' and q be the quotient and the remainder when k is divided by n; that is, $k = k'n + q$, $0 \le q < n$, and

$$0 \le k' < \frac{k}{n} < \frac{n!}{n} = (n-1)!.$$

From the inductive hypothesis, there are integers $d_1' < d_2' < \cdots < d_s'$, $s \le n-1$, such that $d_i' \mid (n-1)!$, $i = 1, 2, \ldots, s$ and $kk' = d_1' + d_2' + \cdots + d_s'$. Hence $k = nd_1' + nd_2' + \cdots + nd_s' + q$. If $q = 0$, then $k = d_1 + d_2 + \cdots + d_s$, where $d_i = nd_i'$, $i = 1, 2, \ldots, s$, are distinct divisors of $n!$.

If $q \neq 0$, then $k = d_1 + d_2 + \cdots + d_{s+1}$, where $d_i = nd_i'$, $i = 1, 2, \ldots, s$, and $d_{s+1} = q$. It is clear that $d_i \mid n!$, $i = 1, 2, \ldots, s$, and $d_{s+1} \mid n!$, since $q < n$. On the other hand, $d_{s+1} < d_1 < d_2 < \cdots < d_s$, because $d_{s+1} = q < n \le nd_1' = d_1$. Therefore k can be written as a sum of at most n distinct divisors of $n!$, as claimed.

47. Let $n > 1$ be an odd integer. Prove that n does not divide $3^n + 1$.

Proof: Assume to the contrary that there is a positive odd integer n that divides $3^n + 1$.

Let p be the smallest prime divisor of n. Then p divides $3^n + 1$; that is, $3^n \equiv -1 \pmod{p}$, implying that $3^{2n} \equiv 1 \pmod{p}$. By Fermat's little theorem, we also have $3^{p-1} \equiv 1 \pmod{p}$. By Corollary 1.23,

$$3^{\gcd(2n, p-1)} \equiv 1 \pmod{p}.$$

Because p is the smallest prime divisor of n, $\gcd(n, p-1) = 1$. Because n is odd, $p - 1$ is even. Hence $\gcd(2n, p-1) = 2$. It follows that $3^2 \equiv 1 \pmod{p}$, or p divides 8, which is impossible (since p is odd).

48. Let a and b be positive integers. Prove that the number of solutions (x, y, z) in nonnegative integers to the equation $ax + by + z = ab$ is

$$\frac{1}{2}[(a+1)(b+1) + \gcd(a, b) + 1].$$

Proof: It is clear that for each solution (x, y, z) in nonnegative integers to $ax + by + z = ab$ we have the solution (x, y) in nonnegative integers to $ax + by \le ab$. Conversely, for each solution (x, y) to $ax + by \le ab$ we have the solution $(x, y, ab - ax - by)$ to the given equation. Hence it suffices to count the number of solutions (x, y) in nonnegative integers to $ax + by \le ab$.

Clearly, these solutions correspond to points of integer coordinates in the rectangle $[0, b] \times [0, a]$.

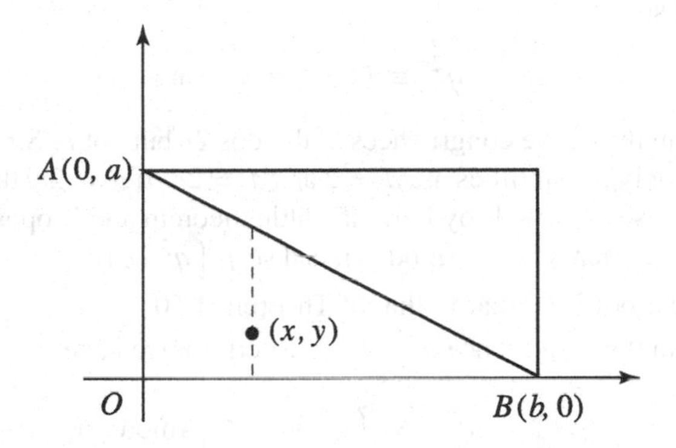

The number of lattice points (that is, points with integer coordinates) in this rectangle is $(a+1)(b+1)$. The condition $ax+by \le ab$ means that the point (x, y) is situated under or on the diagonal AB. Because of the symmetry, the desired number of points (x, y) is

$$\frac{1}{2}(a+1)(b+1) + \frac{d}{2},$$

where d is the number of such points on the diagonal AB. In order to find d, note that $ax + by = ab$ is equivalent to $y = a - \frac{a}{b}x$. The number of integers in the array

$$\frac{1 \cdot a}{b}, \frac{2 \cdot a}{b}, \ldots, \frac{b \cdot a}{b}$$

is $\gcd(a, b)$. We also need to count the point $A(0, a)$; hence $d = \gcd(a, b) + 1$ and the conclusion follows.

49. Order! Order, please!

(1) Let p be an odd prime, and let q and r be primes such that p divides $q^r + 1$. Prove that either $2r \mid p - 1$ or $p \mid q^2 - 1$.

(2) Let $a > 1$ and n be given positive integers. If p is a prime divisor of $a^{2^n} + 1$, prove that $p - 1$ is divisible by 2^{n+1}.

Proof: In this problem, we apply Propositions 1.30 repeatedly.

(1) Let $d = \text{ord}_p(q)$ (the order of q modulo p). Since $p \mid q^r + 1$ and $p > 2$, we have

$$q^r \equiv -1 \not\equiv 1 \pmod{p},$$

and so

$$q^{2r} \equiv (-1)^2 \equiv 1 \pmod{p}.$$

From the above congruences, d divides $2r$ but not r. Since r is prime, the only possibilities are $d = 2$ and $d = 2r$. If $d = 2r$, then $2r \mid p-1$ because $d \mid p-1$, by Fermat's little theorem and Proposition 1.30. If $d = 2$, then $q^2 \equiv 1 \pmod{p}$, and so $p \mid q^2 - 1$.

(2) The proof is similar to that of Theorem 1.50.
From the congruence $a^{2^n} \equiv -1 \pmod{p}$, we have

$$a^{2^{n+1}} = \left(a^{2^n}\right)^2 \equiv 1 \pmod{p}.$$

By Proposition 1.30, $\text{ord}_p(a)$ divides 2^{n+1}. Since $2^{2^n} \equiv -1 \pmod{p}$, $\text{ord}_p(a) = 2^{n+1}$. Clearly, $\gcd(a, p) = 1$. By Fermat's little theorem, we have $a^{p-1} \equiv 1 \pmod{p}$. By Proposition 1.30, we conclude that 2^{n+1} divides $p - 1$.

Note: Setting $a = 2$ in (2) shows that if p is a prime divisor of the Fermat number f_n, then $p - 1$ is divisible by 2^{n+1}.

50. [APMO 2004] Prove that

$$\left\lfloor \frac{(n-1)!}{n(n+1)} \right\rfloor$$

is even for every positive integer n.

Proof: One checks directly that the conclusion holds for $n = 1, 2, \ldots, 6$. Now we assume that $n \geq 6$. We consider three cases.

In the first case, we assume that $n = p$ is prime. Then $n + 1 = p + 1$ is even. Hence $n + 1 = 2 \cdot \frac{p+1}{2}$ divides $(n-1)! = (p-1)!$ and

$$k = \frac{(n-1)!}{n+1} = \frac{(p-1)!}{p+1}$$

is an even integer, and so $k+1$ is an odd integer. By Wilson's theorem,

$$k+1 \equiv \frac{(p-1)!}{p+1} + 1 \equiv \frac{-1}{1} + 1 \equiv 0 \pmod{p},$$

implying that $\frac{k+1}{p}$ is an odd integer; that is,

$$\frac{\frac{(p-1)!}{p+1} + 1}{p} = \frac{(p-1)!}{p(p+1)} + \frac{1}{p}$$

is an odd integer. Hence

$$\left\lfloor \frac{(n-1)!}{n(n+1)} \right\rfloor = \left\lfloor \frac{(p-1)!}{p(p+1)} \right\rfloor$$

is even.

In the second case, we assume that $n+1 = p$ is prime. Then $n = p-1$ is even. Hence $n = 2 \cdot \frac{p-1}{2}$ divides $(n-1)! = (p-2)!$ and

$$k' = \frac{(n-1)!}{n} = \frac{(p-2)!}{p-1}$$

is an even integer, and so $k'+1$ is an odd integer. By Wilson's theorem,

$$k'+1 \equiv \frac{(p-2)!}{p-1} + 1 \equiv \frac{(p-1)!}{(p-1)^2} + 1 \equiv -1 + 1 \equiv 0 \pmod{p},$$

implying that $\frac{k'+1}{p}$ is an odd integer; that is,

$$\frac{\frac{(p-2)!}{p-1} + 1}{p} = \frac{(p-2)!}{p(p-1)} + \frac{1}{p}$$

is an odd integer. Hence

$$\left\lfloor \frac{(n-1)!}{n(n+1)} \right\rfloor = \left\lfloor \frac{(p-2)!}{p(p-1)} \right\rfloor$$

is even.

In the third case, we assume that both n and $n+1$ are composite. It is not difficult to show that both n and $n+1$ divide $(n-1)!$. Since $\gcd(n, n+1) = 1$, we conclude that $n(n+1)$ divides $(n-1)!$; that is,

$$\frac{(n-1)!}{n(n+1)}$$

is an integer. Also, using Legendre's function, it is easy to see that this integer is also even.

51. [ARML 2002] Determine all the positive integers m each of which satisfies the following property: there exists a unique positive integer n such that there exist rectangles that can be divided into n congruent squares and also into $n+m$ congruent squares.

Solution: The integer m satisfies the conditions of the problem if and only if m is in the set

$$S = \{8,\ p, 2p, \text{ and } 4p, \text{ where } p \text{ is an odd prime}\}.$$

Without loss of generality, consider a rectangle $ABCD$ that can be divided into $n+m$ squares of side 1 and m larger squares of side x. Because the sides of $ABCD$ have integer lengths, x must be a rational number. Write $x = \frac{a}{b}$, where a and b are relatively prime integers. Because $x > 1$, $a > b$. The area of $ABCD$ is

$$(n+m)\cdot 1 = n \cdot \left(\frac{a}{b}\right)^2.$$

Solving the above equation for n gives

$$n = \frac{mb^2}{a^2 - b^2} = \frac{mb^2}{(a+b)(a-b)}.$$

Because $\gcd(a, b) = 1$, $\gcd(b, a+b) = \gcd(a, a-b) = 1$, and so $(a+b)(a-b)$ divides m. Note also that $a+b$ and $a-b$ have the same parity.

If m has two odd factors each of which is greater than 1, write $m = ijk$, where $j > 1$ and $k > 1$ are odd integers. Then $(a+b, a-b) = (j, k)$ and $(a+b, a-b) = (jk, 1)$ lead to two distinct values for n, namely, $n = \frac{i(j-k)^2}{4}$ and $n = \frac{i(jk-1)^2}{4}$, contradicting the uniqueness of n. Hence m has at most one odd factor greater than 1; that is, $m = 2^c$ or $2^c \cdot p$ for some prime p. We consider these two cases separately.

In the first case, we assume that $m = 2^c$. It is not difficult to check that there is no solution for n when $c = 1$ and 2. If $c > 3$, then $(a-b, a+b) = (2, 4)$ and $(a+b, a-b) = (2, 8)$ lead to two distinct values for n, namely, $n = 2^{c-3}$ and $n = 2^{c-4}$, contradicting the uniqueness of n. For $c = 3$ (and $m = 8$), we must have $(a, b) = (2, 4)$ and $n = 1$.

In the second case, we assume that $m = 2^c \cdot p$. Similar to the first case, we can show that $c \le 2$ (by also considering $(a+b, a-b) = (1, p)$). Hence

$m = p, 2p$, or $4p$. The following table shows that all these values work.

m	$(a+b, a-b)$	(a, b)	n
p	$(p, 1)$	$\left(\frac{p+1}{2}, \frac{p-1}{2}\right)$	$\frac{(p-1)^2}{4}$
$2p$	$(p, 1)$	$\left(\frac{p+1}{2}, \frac{p-1}{2}\right)$	$\frac{(p-1)^2}{2}$
$4p$	$(p, 1)$ or $(2p, 2)$	$\left(\frac{p+1}{2}, \frac{p-1}{2}\right)$ or $(p+1, p-1)$	$(p-1)^2$

52. Determine all positive integers n such that n has a multiple whose digits are nonzero.

Solution: We claim that an integer n satisfies the conditions of the problem if and only if n is not a positive multiple of 10. We call a number *good* if it satisfies the conditions of the problem. Clearly, multiples of 10 are not good since their multiples always end in the digit 0. We will show that all the other positive integers are good. Let n be a positive integer not divisible by 10. We consider a few cases.

In the first case, we assume that $n = 5^k$ or $n = 2^k$ for some positive integer k. As we have shown in Example 1.49, there exist k-digit multiples of n whose digits are nonzero, implying that n is good.

In the second case, we assume that n is relatively prime to 10. We claim that n has a multiple whose digits are all equal to 1. We take

$$\underbrace{11\ldots1}_{\varphi(9n)} = \frac{10^{\varphi(9n)} - 1}{9},$$

which is divisible by n, from Euler's theorem.

In the third case, we assume that $n = a^s \cdot m$, where $a = 2$ or 5 and m is relatively prime to 10. As we have discussed in the first case, there is an s-digit multiple of a^s whose digits are nonzero. Let $t = \overline{a_{s-1}a_{s-2}\ldots a_0}$ be that number. We consider the sequence

$$\overline{a_{s-1}a_{s-2}\ldots a_0},\ \overline{a_{s-1}a_{s-2}\ldots a_0 a_{s-1}a_{s-2}\ldots a_0},\ \ldots;$$

that is, the kth number in the sequence is the concatenation (the number obtained by writing them one after another) of k t's. As we have shown in the second case, two terms, say the ith and jth terms ($i < j$) in the sequence, are congruent to each other modulo m. It follows that

$$\underbrace{\overline{a_{s-1}a_{s-2}\ldots a_0\ldots a_{s-1}a_{s-2}\ldots a_0}}_{j-i\ \overline{a_{s-1}a_{s-2}\ldots a_0}\text{'s}}\underbrace{00\ldots0}_{(j-i)s\ 0\text{'s}} \equiv 0 \pmod{m}.$$

Since $\gcd(m, 10) = \gcd(m, a^s) = 1$ and a^s divides $t = \overline{a_{s-1}a_{s-2}\ldots a_0}$, it follows that

$$\underbrace{\overline{a_{s-1}a_{s-2}\ldots a_0\ldots a_{s-1}a_{s-2}\ldots a_0}}_{j-i\ \overline{a_{s-1}a_{s-2}\ldots a_0}\text{'s}}$$

is a multiple of $n = a^s \cdot m$ whose digits are nonzero.

5

Solutions to Advanced Problems

1. [MOSP 1998]

(a) Prove that the sum of the squares of 3, 4, 5, or 6 consecutive integers is not a perfect square.

(b) Give an example of 11 consecutive positive integers the sum of whose squares is a perfect square.

Proof: Define $s(n,k) = n^2 + (n-1)^2 + \cdots + (n+k-1)^2$ as the sum of squares of k consecutive integers, the least of which is n.

(1) Note that $s(n-1,3) = (n-1)^2 + n^2 + (n+1)^2 = 3n^2 + 2$. Since $s(n-1,3) \equiv 2 \pmod 3$, $s(n-1,3)$ is not a perfect square; that is, the sum of the squares of 3 consecutive integers is not a perfect square.

Note that $s(n,4) = 4(n^2+3n+3)+2$. Since $s(n,4) \equiv 2 \pmod 4$, $s(n,4)$ is not a perfect square; that is, the sum of the squares of 4 consecutive integers is not a perfect square.

Note that $s(n-2,5) = 5(n^2+2)$. Since $s(n-2,5) \equiv n^2+2 \equiv 2$ or 3 modulo 4, $s(n-2,5)$ is not a perfect square; that is, the sum of the squares of 5 consecutive integers is not a perfect square.

Note that $s(n-2,6) = 6n^2+6n+19$. Since $n^2+n = n(n+1)$ is even, $s(n-2,6) \equiv 6n(n+1)+19 \equiv 3 \pmod 4$, and so $s(n-2,6)$ is not a perfect square; that is, the sum of the squares of 6 consecutive integers is not a perfect square.

(2) We have $s(n-5,11) = 11(n^2+10)$. It remains to find n such that $11(n^2+10)$ is a perfect square. Hence 11 must divide n^2+10, or $n^2-1 \equiv n^2+10 \pmod{11}$. Consequently, $n-1 \equiv 0 \pmod{11}$ or $n+1 \equiv 0 \pmod{11}$, or $n = 11m \pm 1$ for some integer k. It follows that $s(n-5,11) = 11[(11m\pm 1)^2+10] = 11^2(11m^2 \pm 2m + 1) =$

$11^2[10m^2+(m\pm1)^2]$. We observe that for $m=2$, $10m^2+(m+1)^2 = 49 = 7^2$, which leads to an example so $s(18, 11) = 77^2$.

2. [MOSP 1998] Let $S(x)$ be the sum of the digits of the positive integer x in its decimal representation.

   (a) Prove that for every positive integer x, $\frac{S(x)}{S(2x)} \le 5$. Can this bound be improved?

   (b) Prove that $\frac{S(x)}{S(3x)}$ is not bounded.

Proof:

(a) The maximum carry is 1. This implies that the only carries in $2x$ are the ones accounted for in $S(2d)$ for each digit d in the decimal representation of x. Hence $S(2x) = \sum S(2d)$, where the sum is taken over all the digits of x. It is clear that $S(d)/S(2d) \le 5$ for every decimal digit $d \ne 0$. Thus

$$\frac{S(x)}{S(2x)} = \frac{\sum S(d)}{\sum S(2d)} \le 5.$$

This bound cannot be improved, since $S(5) = 5S(10)$.
One can also apply Proposition 1.45 (d) to obtain $S(x) = S(10x) \le S(5)S(2x) = 5S(2x)$.

(b) Let

$$p_k = \underbrace{33\ldots3}_{k}4.$$

Then

$$3p_k = 3(\underbrace{33\ldots3}_{k+1}+1) = \underbrace{99\ldots9}_{k+1}+3 = 1\underbrace{00\ldots0}_{k}2.$$

Thus

$$\frac{S(p_k)}{S(3p_k)} = \frac{3k+4}{3},$$

which is unbounded. This completes our proof.

3. Most positive integers can be expressed as a sum of two or more consecutive positive integers. For example, $24 = 7 + 8 + 9$ and $51 = 25 + 26$. A positive integer that cannot be expressed as a sum of two or more consecutive positive integers is therefore *interesting*. What are all the interesting integers?

Solution: A number n is interesting if and only if n is power of 2; that is, $n = 2^k$ for some nonnegative integer k.

Assume that n is not interesting. We can write

$$n = m + (m + 1) + \cdots + (m + k) = \frac{(k + 1)(2m + k)}{2} \qquad (*)$$

for some positive integers m and k. Since $k + 1$ and $2m + k$ are of different parity, one of them is an odd integer greater than 3, and so n must have an odd divisor greater than 3. It follows that 2^k (for every positive integer k) is interesting.

It remains to show that all other positive integers n are not interesting. We write $n = 2^h \cdot \ell$ where h is nonnegative and ℓ is an odd number greater than 1. (Note that $2^{h+1} \neq \ell$.) If $2^{h+1} < \ell$, n is not interesting since we can set

$$k = 2^{h+1} - 1 \quad \text{and} \quad m = \frac{\ell - k}{2} = \frac{\ell + 1 - 2^{h+1}}{2}$$

in $(*)$; if $2^{h+1} > \ell$, n is not interesting since we can set

$$k = \ell - 1 \quad \text{and} \quad m = \frac{2^{h+1} - k}{2} = \frac{2^{h+1} + 1 - \ell}{2}$$

in $(*)$. Hence, in any case, n is not interesting if n has a odd divisor greater than 1, completing our proof.

4. Set $S = \{105, 106, \ldots, 210\}$. Determine the minimum value of n such that any n-element subset T of S contains at least two non-relatively prime elements.

Solution: The minimum value of n is 26.

Our first step is to compute the number of prime numbers in the set S. For any positive integer k, let A_k denote the subset of multiples of k in S, and let $P = \{2, 3, 5, 7, 11\}$. We compute the cardinality of the subset of S consisting of numbers divisible by one or more of 2, 3, 5, 7, or 11:

$$A = \bigcup_{k \in P} A_k = A_2 \cup A_3 \cup A_5 \cup A_7 \cup A_{11},$$

using the inclusion and exclusion principle, as follows:

$$\begin{aligned}|A| &= \sum_{k\in P}|A_k| - \sum_{i<j\in P}|A_i\cap A_j| + \sum_{i<j<k\in P}|A_i\cap A_j\cap A_k| \\ &\quad - \sum_{i<j<k<l\in P}|A_i\cap A_j\cap A_k\cap A_l| + \left|\bigcap_{k\in P}A_k\right| \\ &= 137 - 66 + 16 - 1 + 0 \\ &= 86.\end{aligned}$$

We now see that the only composite number in S that is not in A is $13^2 = 169$, since $13 \cdot 17 = 221 > 210$. Therefore S consists of 87 composite numbers and 19 primes.

We can now prove that given any 26 numbers in S, there exist two of them that are not relatively prime. By the pigeonhole principle, since there are 19 primes in S, at least 7 of the 26 numbers we have chosen are composite, which means that at least 6 of the numbers are in A. But this means, again by the pigeonhole principle, that two of them belong to the same set A_k, for some $k \in P$. Thus they share a common factor (namely k), and hence are not relatively prime.

Finally, we can construct a subset of S with 25 elements in which every pair of elements is relatively prime. Let P denote the set of all primes in S; then we can see that the set

$$\begin{aligned}&P \cup \{11^2,\ 5^3,\ 2^7,\ 3^2\cdot 17,\ 13^2,\ 7\cdot 29\} \\ &\qquad = P \cup \{121, 125, 128, 153, 169, 203\}\end{aligned}$$

is a set of 25 numbers that are all mutually relatively prime.

5. [St. Petersburg 1997] The number

$$\underbrace{99\ldots99}_{1997\ 9\text{'s}}$$

is written on a blackboard. Each minute, one number written on the blackboard is factored into two factors and erased, each factor is (independently) increased or diminished by 2, and the resulting two numbers are written. Is it possible that at some point (after the first minute) all of the numbers on the blackboard equal 9?

Solution: The answer is negative.

Adding or subtracting 2 from a number motivates us to consider arithmetic modulo 4. Since $a + 2 \equiv a - 2 \pmod 4$ for any integer a, adding or subtracting 2 becomes the same operation modulo 4. Note first that the original number is congruent to 3 modulo 4. We claim that there is always a number congruent to 3 modulo 4: factoring such a number gives one factor congruent to 1 modulo 4, and changing that by 2 in either direction gives a number congruent to 3 modulo 4. On the other hand, 9 is congruent to 1 modulo 4, and so we cannot have all 9's written on the board at any moment.

6. IMO 1986] Let d be any positive integer not equal to 2, 5, or 13. Show that one can find distinct a, b in the set $\{2, 5, 13, d\}$ such that $ab - 1$ is not a perfect square.

First Proof: Since $2 \cdot 5 - 1 = 3^2$, $2 \cdot 13 - 1 = 5^2$, and $5 \cdot 13 - 1 = 8^2$, we will look for a non-perfect square in the set $\{2d - 1, 5d - 1, 13d - 1\}$. Assume to the contrary that all these numbers are perfect squares; that is,

$$2d - 1 = a^2, \quad 5d - 1 = b^2, \quad \text{and} \quad 13d - 1 = c^2,$$

where a, b, and c are integers. Then a is an odd number, say, $a = 2x + 1$ and $d = 2x(x+1)+1$. Since $x(x+1)$ is always even, it follows that $d \equiv 1 \pmod 4$, and so b and c are even. Assume that $b = 2y$ and $c = 2z$. From $5d = b^2 + 1$ and $13d = c^2 + 1$, we have $8d = c^2 - b^2$. Thus,

$$d = \frac{4y^2 + 1}{5} = \frac{4z^2 + 1}{13} = \frac{4z^2 - 4y^2}{8} = \frac{z^2 - y^2}{2}.$$

It follows that z and y are of equal parity. In this case, $z^2 - y^2 \equiv 0 \pmod 4$, while $d \equiv 1 \pmod 4$. Thus, we get a contradiction.

Second Proof: We operate modulo 16. We first calculate n^2 modulo 16 for $n = 0, 1, \ldots, 7, 8$ to see that the possible residues modulo 16 are

$$0, \ 1, \ 4, \ 9.$$

If $2d - 1$ is not a perfect square, we are done. Assume that $2d - 1$ is a perfect square. Then $2d - 1$ is congruent to 0, 1, 4, or 9 modulo 16. Since $2d$ is even, $2d$ is congruent to 2 or 10 modulo 16, implying that d is congruent to

1, 5, 9, or 13 modulo 16. Thus, we have the following table (modulo 16):

d	$5d-1$	$13d-1$
1	4	**12**
5	**8**	4
9	**12**	4
13	4	**8**

Note that all the boldfaced numbers are not perfect squares, and there is such a number in each row. Thus, for all possible values of d that make $2d-1$ a perfect square, at least one of $5d-1$ and $13d-1$ is a not perfect square, and we are done.

7. [Russia 2001] A heap of balls consists of one thousand 10-gram balls and one thousand 9.9-gram balls. We wish to pick out two heaps of balls with equal numbers of balls in them but different total weights. What is the minimal number of weighings needed to do this? (The balance scale reports the weight of the objects in the left pan minus the weight of the objects in the right pan.)

   Proof: It is clear that one has to use at least one weighing. We claim that it is also enough.

   Split the two thousand balls into three heaps H_1, H_2, H_3 of 667, 667, and 666 balls, respectively. Weigh heaps H_1 and H_2 against each other. If the total weights are not equal, we are done. Otherwise, discard one ball from H_1 to form a new heap H_1' of 666 balls. We claim that H_1' and H_3 have different weights. If not, then they have the same number of 10-gram balls, say, n. Then H_1 and H_2 either each had n 10-gram balls or each had $n+1$ 10-gram balls. This would imply that 1000 equals $3n$ or $3n+2$, which is impossible.

8. [China 2001] We are given three integers a, b, and c such that a, b, c, $a+b-c$, $a+c-b$, $b+c-a$, and $a+b+c$ are seven distinct primes. Let d be the difference between the largest and smallest of these seven primes. Suppose that 800 is an element in the set $\{a+b, b+c, c+a\}$. Determine the maximum possible value of d.

   First Solution: The answer is 1594.

   First, observe that a, b, and c must all be odd primes; this follows from the assumption that the seven quantities listed are distinct primes and the fact that there is only one even prime, 2. (If, say, a is even, then b and c must be odd. Then $a+b-c$ and $a+c-b$ must both be even, and so equal to 2.) Therefore, the smallest of the seven primes is at least 3.

Second, assume without loss of generality that $a+b=800$. Because $a+b-c>0$, we must have $c<800$. We also know that c is prime; therefore, since $799=17\cdot 47$, we have $c\le 797$. It follows that the largest prime, $a+b+c$, is no more than 1597. Combining these two bounds, we can bound d by $d\le 1597-3=1594$.

It remains to observe that we can choose $a=13$, $b=787$, $c=797$ to achieve this bound. The other four primes are then 3, 23, 1571, and 1597.

Second Solution: assume without loss of generality that $a+b=800$. (Clearly, both a and b are odd.) Then c, $a+b+c=800+c$, and $a+b-c=800-c$ are primes. Consider c, $800+c$, and $800-c$ modulo 3. It is not difficult to see that exactly one of them is congruent to 0 modulo 3; that is, one of them is equal to 3. Consequently, we have either $c=3$ or $800-c=3$ (and $c=797$). If $c=3$, $d<a+b+c=803$. If $c=797$, then $d\le a+b+c-3=1594$. We can finish as we did in the first solution.

9. Prove that the sum

$$S(m,n)=\frac{1}{m}+\frac{1}{m+1}+\cdots+\frac{1}{m+n}$$

is not an integer for any given positive integers m and n.

Proof: Assume to the contrary that $S(m,n)$ is an integer for some positive integers m and n. Clearly, $n\ge 1$. Consequently, there are even integers among the numbers $m, m+1, \ldots, m+n$. Let ℓ denote lcm$(m, m+1, \ldots, m+n)$. Then ℓ is even. We have

$$\ell S(m,n)=\frac{\ell}{m}+\frac{\ell}{m+1}+\cdots+\frac{\ell}{m+n}. \qquad (*)$$

By our assumption, the left-hand side of the above identity is even. We will reach a contradiction by showing that the right-hand side is odd.

For every integer i with $0\le i\le n$, assume that 2^{a_i} fully divides $m+i$. Let $m=\max\{a_0, a_1, \ldots, a_n\}$. It follows that $2^m \,\|\, \ell$.

Assume that $a_j=m$ (where $0\le j\le n$). We claim that j is unique. Assume to the contrary that $a_j=a_{j_1}$ with $0\le j<j_1\le n$. Then $m+j=2^{a_j}\cdot k$ and $m+j_1=2^{a_{j_1}}\cdot k_1$, where k and k_1 are odd positive integers. Hence $k+1$ is an even integer between k and k_1, and so

$$m+j<2^{a_j}\cdot(k+1)<2^{a_j}\cdot k_1=2^{a_{j_1}}\cdot k_1=m+j_1,$$

implying that 2^{a_j+1} divides $2^{a_j} \cdot (k+1)$, contradicting the maximality of $a_j = m$. Thus, such a j is unique. It follows that $\frac{\ell}{m+i}$ is an even integer for all $0 \le i \le n$ with $i \ne j$, and $\frac{\ell}{m+j}$ is odd. Hence all but one summand on the right-hand side of $(*)$ are even, implying that the right-hand side of $(*)$ is odd, contradicting the fact that the left-hand side of $(*)$ is even. Hence our original assumption was wrong and $S(m,n)$ is not an integer.

10. [St. Petersburg 2001] For all positive integers $m > n$, prove that

$$\operatorname{lcm}(m,n) + \operatorname{lcm}(m+1,n+1) > \frac{2mn}{\sqrt{m-n}}.$$

Proof: Let $m = n + k$. Then

$$\begin{aligned}\operatorname{lcm}(m,n) &+ \operatorname{lcm}(m+1,n+1)\\ &= \frac{mn}{\gcd(m,n)} + \frac{(m+1)(n+1)}{\gcd(m+1,n+1)}\\ &> \frac{mn}{\gcd(n+k,n)} + \frac{mn}{\gcd(m+1,n+1)}\\ &= \frac{mn}{\gcd(k,n)} + \frac{mn}{\gcd(n+k+1,n+1)}\\ &= \frac{mn}{\gcd(k,n)} + \frac{mn}{\gcd(k,n+1)}.\end{aligned}$$

Now, $\gcd(k,n) \mid k$, and $\gcd(k,n+1) \mid k$. We conclude that $\gcd(k,n)$ has no common prime factor with $\gcd(k,n+1)$, because if it did, $n+1$ would have a common prime factor with n, which is impossible. Since both divide k, so does their product, implying that $\gcd(k,n)\gcd(k,n+1) \le k$. Consequently,

$$\begin{aligned}\operatorname{lcm}(m,n) + \operatorname{lcm}(m+1,n+1) &> \frac{mn}{\gcd(k,n)} + \frac{mn}{\gcd(k,n+1)}\\ &\ge 2mn\sqrt{\frac{1}{\gcd(k,n)\gcd(k,n+1)}} \ge 2mn\sqrt{\frac{1}{k}} = \frac{2mn}{\sqrt{m-n}}\end{aligned}$$

by the **AM-GM inequality**.

11. Prove that each nonnegative integer can be represented in the form $a^2 + b^2 - c^2$, where a, b, and c are positive integers with $a < b < c$.

First Proof: Let k be a nonnegative integer.

If k is even, say $k = 2n$, the conclusion follows from the identity

$$2n = (3n)^2 + (4n-1)^2 - (5n-1)^2$$

and the simple algebraic facts $3n < 4n - 1 < 5n - 1$ for $n > 1$,

$$0 = 3^2 + 4^2 - 5^2, \quad \text{and} \quad 2 = 5^2 + 11^2 - 12^2.$$

If k is odd, we use the identity

$$2n + 3 = (3n+2)^2 + (4n)^2 - (5n+1)^2,$$

where for $n > 2$, $3n + 2 < 4n < 5n + 1$. Since

$$1 = 4^2 + 7^2 - 8^2,\ 3 = 4^2 + 6^2 - 7^2,$$
$$5 = 4^2 + 5^2 - 6^2, \text{ and } 7 = 6^2 + 14^2 - 15^2,$$

we have exhausted the case k odd as well.

Second Proof: We present a more general approach for this problem. The key fact is that the positive differences between consecutive perfect squares are linearly increasing. For every nonnegative integer k, we choose a large positive integer a with different parity from that of k. We then set $c = b+1$. Then $k = a^2 + b^2 - c^2 = a^2 - (2b+1)$. Since k and a has different parity, $a^2 - k$ is odd, and so

$$b = \frac{a^2 - k - 1}{2}$$

is a positive integer. Since the left-hand side of the above equation is a quadratic in a, its value is greater than a for large a, and so the condition $a < b < c = b + 1$ is satisfied, and we are done.

12. Determine whether there exists a sequence of strictly increasing positive integers $\{a_k\}_{k=1}^{\infty}$ such that the sequence $\{a_k + a\}_{k=1}^{\infty}$ contains only finitely many primes for all integers a.

Note: One easily thinks about $a_k = k!$. But it is then difficult to deal with $a = 1$ or $a = -1$. We present two ways to modify this sequence.

First Solution: The answer is positive. There exists such a sequence. Indeed, for every positive integer k, let $a_k = (k!)^3$. If $a = \pm 1$, then $a_k + a = (k!)^3 \pm 1$ is composite, since polynomials $x^3 + 1$ and $x^3 - 1$ factor into $(x+1)(x^2 - x + 1)$ and $(x-1)(x^2 + x + 1)$, respectively. If $|a| > 1$, then a divides $k!$ for $k \geq |a|$, implying that a also divides $a_k + a$ for $k \geq |a|$.

Second Solution: (By Kevin Modzelewski) We set $a_k = (2k)! + k$. For all integers $k \geq |a|$ and $k \geq 2 - a$, we have $2 \leq k + a \leq 2k$. Therefore, $a_k + a$ is divisible by $k + a$, and thus composite, for all such k.

13. Prove that for different choices of signs $+$ and $-$ the expression

$$\pm 1 \pm 2 \pm 3 \pm \cdots \pm (4n+1)$$

yields all odd positive integers less than or equal to $(2n+1)(4n+1)$.

Proof: We induct on n. For $n = 1$, from $\pm 1 \pm 2 \pm 3 \pm 4 \pm 5$ we obtain all odd positive integers less than or equal to $(2+1)(4+1) = 15$:

$$\begin{aligned} &+1-2+3+4-5=1, && -1+2+3+4-5=3,\\ &-1+2+3-4+5=5, && -1+2-3+4+5=7,\\ &-1-2+3+4+5=9, && +1-2+3+4+5=11,\\ &-1+2+3+4+5=13, && +1+2+3+4+5=15. \end{aligned}$$

Assume that the conclusion is true for $n = k$, where k is some positive integer; that is, from $\pm 1 \pm 2 \pm \cdots \pm (4k+1)$ with suitable choices of signs $+$ and $-$ we obtain all odd positive integers less than or equal to $(2k+1)(4k+1)$. Now we assume that $n = k + 1$.

Observe that $-(4k+2) + (4k+3) + (4k+4) - (4k+5) = 0$. Hence from $\pm 1 \pm 2 \pm \cdots \pm (4k+5)$ for suitable choices of signs + and $-$ we obtain all odd positive numbers less than or equal to $(2k+1)(4k+1)$.

It suffices to obtain all odd integers m such that

$$(2k+1)(4k+1) < m \leq (2k+3)(4k+5) = (2n+1)(4n+1). \quad (*)$$

There are

$$\frac{(2k+3)(4k+5) - (2k+1)(4k+1)}{2} = 8k+7$$

such odd integers m. Each of these integers can be written in exactly one of the following forms:

$$(2n+3)(4n+5) = +1 + 2 + \cdots + (4n+5),$$

or

$$(2n+3)(4n+5) - 2k$$
$$= +1 + 2 + \cdots + (k-1) - k + (k+1) + \cdots + (4n+4) + (4n+5),$$

for $k = 1, 2, \ldots, 4n+5$, or

$$(2n+1)(4n+5) - 2\ell$$
$$= +1 + 2 + \cdots + (\ell-1) - \ell + (\ell+1) + \cdots + (4n+4) - (4n+5),$$

for $\ell = 1, 2, \ldots, 4n+1$. Hence all numbers m from $(*)$ are obtained, completing the inductive step.

14. Let a and b be relatively prime positive integers. Show that

$$ax + by = n$$

has nonnegative integer solutions (x, y) for integers $n > ab - a - b$. What if $n = ab - a - b$?

First Proof: We call an integer n *representable* if there are nonnegative integers x and y such that $n = ax + by$.

First we show that $n = ab - a - b$ is not representable. Assume to the contrary that $ab - a - b = ax + by$, where x and y are nonnegative integers. Taking the last equation modulo a and then modulo b leads to $-b \equiv by \pmod{a}$ and $-a \equiv ax \pmod{a}$. Since $\gcd(a, b) = 1$, it follows that $y \equiv -1 \pmod{a}$ and $x \equiv -1 \pmod{b}$. Since x and y are nonnegative, $y \geq a - 1$ and $x \geq b - 1$. Hence

$$ab - a - b = n = ax + by \geq a(b-1) + b(a-1) = 2ab - a - b,$$

which is impossible for positive integers a and b. Therefore, our assumption was wrong and $n = ab - a - b$ is not representable.

Second we show that $n > ab - a - b$ is representable. Since $\gcd(a, b) = 1$, by Proposition 1.24,

$$\{n, n-b, n-2b, \ldots, n-(a-1)b\}$$

is a complete set of residue classes modulo a. Hence there exists exactly one y, with $0 \leq y \leq a-1$, such that $n - yb \equiv 0 \pmod{a}$, or $n - yb = ax$ for some integer x. If $x \geq 0$, we are done. If $x < 0$, then $x \leq -1$, and so

$$n - (a-1)b \leq n - yb = ax \leq -a,$$

or $n \le ab - a - b$, contradicts the condition $n > ab - a - b$. Hence both x and y are nonnegative, and so $n > ab - a - b$ is representable.

Second Proof: We prove the following claim:

> If m and n are integers with $m + n = ab - a - b$, then exactly one of m and n is representable.

If $n > ab - a - b$, then m must be negative, which is clearly not representable. Hence by our claim, n is representable. If $n = ab - a - b$, then since $m = 0$ is clearly representable (with $x = y = 0$), $n = ab - a - b$ is not representable, again by our claim.

It remains to prove our claim. By Bézout's identity, there exist pairs (x, y) of integers such that $ax+by = n$. Since $ax+by = a(x-bt)+b(y+bt)$, we can always reduce or increase x by a multiple of b. Thus, we can always assume that $0 \le x \le b - 1$. Furthermore, a number $n = ax + by$ is representable if and only if it is representable under the additional condition that $0 \le x \le b - 1$. Assume that

$$n = ax + by \quad \text{and} \quad m = as + bt,$$

where x, y, s, and t are integers with both x and s nonnegative integers less than b; that is, $0 \le x, x \le b - 1$. Then

$$ax + by + as + bt = m + n = ab - a - b,$$

or

$$ab - (x + s + 1)a - (y + t + 1)b = 0. \qquad (*)$$

Since $\gcd(a, b) = 1$, equation $(*)$ indicates that b must divide $x + s + 1$. Note that $1 \le x + s + 1 \le 2b - 1$. Hence $x + s + 1 = b$, and the equations $(*)$ becomes $(y+t+1)b = 0$, or $y+t+1 = 0$. It is easy to see that exactly one of y and t is nonnegative, and exactly one of them is negative; that is, one of them is representable and the other is equal to 1.

Note: Can you generalize this result for three pairwise relatively prime numbers a, b, and c?

15. [China 2003] The sides of a triangle have integer lengths k, m, and n. Assume that $k > m > n$ and

$$\left\{\frac{3^k}{10^4}\right\} = \left\{\frac{3^m}{10^4}\right\} = \left\{\frac{3^n}{10^4}\right\}.$$

Determine the minimum value of the perimeter of the triangle.

Solution: It suffices to find positive integers k, m, and n with $k > m > n$ and $k < m + n$ such that

$$3^k \equiv 3^m \equiv 3^n \pmod{10^4},$$

or

$$3^k \equiv 3^m \equiv 3^n \pmod{2^4} \quad \text{and} \quad 3^k \equiv 3^m \equiv 3^n \pmod{5^4}. \qquad (*)$$

Let $d_1 = \text{ord}_{2^4}(3)$, $d_2 = \text{ord}_{5^4}(3)$, and $d = \gcd(d_1, d_2)$. By Proposition 1.30, d divides both of $k - m$ and $m - n$.

It is easy to check that $d_1 = 4$. We note that d_2 divides $\varphi(5^4) = 5^4 - 5^3 = 500$, by Proposition 1.30 again. We claim that $d_2 = 500$. If $d_2 < 500$, it must be a divisor of either $250 = \frac{500}{2}$ or $100 = \frac{500}{5}$. It suffices to show that

(a) $3^{250} \not\equiv 1 \pmod{5^4}$ and
(b) $3^{100} \not\equiv 1 \pmod{5^4}$.

We establish (a) by noting that $3^{250} \equiv 3^2 \equiv -1 \pmod 5$, since $\varphi(5) = 4$. By the binomial theorem, we have

$$3^{100} \equiv (10 - 1)^{50} \equiv \binom{50}{48} \cdot 10^2 - \binom{50}{49} \cdot 10 + 1 \not\equiv 1 \pmod{5^4},$$

establishing (b). It follows that $d_2 = 500$ and $d = 500$. Condition $(*)$ is satisfied if and only if both of $k - m$ and $m - n$ are multiples $d = 500$.

We set $m = 500s + n$ and $k = 500t + m = 500(s + t) + n$ for positive integers s and t. The perimeter of the triangle is equal to $k + m + n = 500(2s + t) + 3n$. Condition $k < m + n$ now reads $500t < n$. Therefore, the minimum value of the perimeter is equal to $500 \cdot 3 + 3 \cdot 501 = 3003$, obtained when $s = t = 1$ and $n = 501$.

16. [Baltic 1996] Consider the following two-person game. A number of pebbles are lying on a table. Two players make their moves alternately. A move consists in taking off the table x pebbles, where x is the square of any positive integer. The player who is unable to make a move loses. Prove that there are infinitely many initial situations in which the player who goes second has a winning strategy.

Proof: Assume to the contrary that there are only finitely many initial situations in which the player who goes second has a winning strategy. Under our assumption, there exists a positive integer N such that if there are $n > N$ pebbles on the table, the player who goes first at the moment has a winning strategy.

Consider the initial situation with $(N+1)^2 - 1$ pebbles on the table. Let $\mathcal{P}_1$ and $\mathcal{P}_2$ denote the players who go first and second, respectively. By our assumption, $\mathcal{P}_1$ has a winning strategy, which requires $\mathcal{P}_1$ to remove x pebbles on his first move. It is clear that $x \neq N^2$, and $\mathcal{P}_2$ is left with at least $(N+1)^2 - 1 - N^2 = 2N > N$ pebbles to make the first move. By our assumption, at this moment, $\mathcal{P}_2$ has a winning strategy. But it is impossible for both players to have a winning strategy for the same initial situation. Hence our original assumption was wrong and there are infinitely many initial situations in which the player who goes second has a winning strategy.

17. [MOSP 1997] Prove that the sequence $1, 11, 111, \ldots$ contains an infinite subsequence whose terms are pairwise relatively prime.

First Proof: Let x_n denote the nth term in the sequence. Then $x_{n+1} - 10x_n = 1$, implying that $\gcd(x_{n+1}, x_n) = 1$. To prove that there is an infinite subsequence of numbers any of two of which are relatively prime, it suffices to prove that no matter how many terms the subsequence contains, it can always contain at least one more term. To do this, note that x_n divides x_{mn}. Let p be the product (or least common multiple) of all the indices already in the subsequence. Then any number in the subsequence divides x_p. Hence x_{p+1} can be added to the subsequence and we are done.

Second Proof: We maintain the same notation as in the first proof. Note that $x_n = \frac{10^n-1}{9}$. By introductory problem 38 (2), we have

$$\gcd(x_m, x_n) = \frac{\gcd(10^m - 1, 10^n - 1)}{9} = \frac{10^{\gcd(m,n)} - 1}{9} = 1$$

for integers m and n with $\gcd(m,n) = 1$. Hence the subsequence $\{x_p\}$ where p are primes satisfies the conditions of the problem.

Note: Euler's proof of the existence of infinitely many primes reveals the connection between these two proofs.

18. Let m and n be integers greater than 1 such that $\gcd(m, n-1) = \gcd(m,n) = 1$. Prove that the first $m-1$ terms of the sequence $n_1, n_2, \ldots$, where $n_1 = mn + 1$ and $n_{k+1} = n \cdot n_k + 1$, $k \geq 1$, cannot all be primes.

Proof: It is straightforward to show that

$$n_k = n^k m + n^{k-1} + \cdots + n + 1 = n^k m + \frac{n^k - 1}{n - 1}$$

for every positive integer k. Hence

$$n_{\varphi(m)} = n^{\varphi(m)} \cdot m + \frac{n^{\varphi(m)} - 1}{n - 1}.$$

From Euler's theorem, $m \mid (n^{\varphi(m)} - 1)$, and since $\gcd(m, n - 1) = 1$, it follows that

$$m \,\Bigg|\, \frac{n^{\varphi(m)} - 1}{n - 1}.$$

Consequently, m divides $n_{\varphi(m)}$. Because $\varphi(m) \le m - 1$, $n_{\varphi(m)}$ is not a prime, and we are done.

19. [Ireland 1999] Find all positive integers m such that the fourth power of the number of positive divisors of m equals m.

Solution: If the given condition holds for some integer m, then m must be a perfect fourth power and we may write its prime factorization as $m = 2^{4a_2}3^{4a_3}5^{4a_5}7^{4a_7}\ldots$ for nonnegative integers $a_2, a_3, a_5, a_7, \ldots$. The number of positive divisors of m equals

$$(4a_2 + 1)(4a_3 + 1)(4a_5 + 1)(4a_7 + 1)\cdots.$$

This number is odd, so m is odd and $a_2 = 0$. Thus,

$$1 = \frac{4a_3 + 1}{3^{a_3}} \cdot \frac{4a_5 + 1}{5^{a_5}} \cdot \frac{4a_7 + 1}{7^{a_7}} \cdots = x_3 x_5 x_7 \cdots,$$

where we write $x_p = \frac{4a_p+1}{p^{a_p}}$ for each p. We proceed to examine x_p through three cases: $p = 3$, $p = 5$, and $p > 5$.

When $a_3 = 1$, $x_3 = \frac{5}{3}$; when $a_3 = 0$ or 2, $x_3 = 1$. When $a_3 > 2$, by **Bernoulli's inequality** we have

$$3^{a_3} = (8 + 1)^{a_3/2} > 8(a_3/2) + 1 = 4a_3 + 1,$$

so that $x_3 < 1$.

When $a_5 = 0$ or 1, $x_5 = 1$; when $a_5 \ge 2$, by Bernoulli's inequality we have

$$5^{a_5} = (24 + 1)^{\frac{a_5}{2}} \ge 24 \cdot \frac{a_5}{2} + 1 = 12a_5 + 1,$$

so that

$$x_5 \le \frac{4a_5+1}{12a_5+1} \le \frac{9}{25}.$$

Finally, for any $p > 5$, when $a_p = 0$ we have $x_p = 1$; when $a_p = 1$ we have $p^{a_p} = p > 5 = 4a_p + 1$, so that $x_p < 1$; and when $a_p > 1$ then again by Bernoulli's inequality we have

$$p^{a_p} > 5^{a_p} > 12a_p + 1,$$

so that as above, $x_p < \frac{9}{25}$.

Now if $a_3 \neq 1$ then we have $x_p \le 1$ for all p. Because $1 = x_2x_3x_5\cdots$ we must actually have $x_p = 1$ for all p. This means that $a_3 \in \{0, 2\}$, $a_5 \in \{0, 1\}$, and $a_7 = a_{11} = \cdots = 0$. Hence $m = 1^4$, $(3^2)^4$, 5^4, or $(3^2 \cdot 5)^4$.

Otherwise, if $a_3 = 1$ then 3 divides $m = 5^4(4a_5+1)^4(4a_7+1)^4\cdots$. Then for some prime $p' \ge 5$, $3 \mid (4a_{p'} + 1)$, so that $a_{p'} \ge 2$; from above, we have $x_{p'} \le \frac{9}{25}$. Then

$$x_3x_5x_7\cdots \le \frac{5}{3} \cdot \frac{9}{25} < 1,$$

which is a contradiction.

Thus, the only such integers m are 1, 5^4, 3^8, and $3^8 \cdot 5^4$, and it is easily verified that these integers work.

20. [Romania 1999]

    (1) Show that it is possible to choose one number out of any 39 consecutive positive integers having the sum of its digits divisible by 11.

    (2) Find the first 38 consecutive positive integers none of which has the sum of its digits divisible by 11.

Proof: Call an integer *deadly* if its sum of digits is divisible by 11, and let $d(n)$ equal the sum of the digits of a positive integer n. We have the following observations:

(a) If n ends in a 0, then the numbers $n, n+1, \ldots, n+9$ differ only in their units digits, which range from 0 to 9. Hence $d(n), d(n+1), \ldots, d(n+9)$ is an arithmetic progression with common difference 1. Thus if $d(n) \not\equiv 1 \pmod{11}$, then one of these numbers is deadly.

(b) Next suppose that n ends in $k \ge 0$ nines. Then $d(n+1) = d(n) + 1 - 9k$: the last k digits of $n+1$ are 0's instead of 9's, and the next digit to the left is 1 greater than the corresponding digit in n.

(c) Finally, suppose that n ends in a 0 and that $d(n) \equiv d(n+10) \equiv 1 \pmod{11}$. Because $d(n) \equiv 1 \pmod{11}$, we must have $d(n+9) \equiv 10 \pmod{11}$. If $n+9$ ends in k 9's, then we have $2 \equiv d(n+10) - d(n+9) \equiv 1 - 9k \pmod{11}$, implying that $k \equiv 6 \pmod{11}$.

(1) Suppose we had 39 consecutive integers, none of them deadly. One of the first ten must end in a 0: call it n. Because none of $n, n+1, \ldots, n+9$ are deadly, we must have $d(n) \equiv 1 \pmod{11}$, by (a) above. Similarly, $d(n+10) \equiv 1 \pmod{11}$ and $d(n+20) \equiv 1 \pmod{11}$. From (c) above, this implies that both $n+9$ and $n+19$ must end in at least six 9's. This is impossible, because $n+10$ and $n+20$ can't both be multiples of one million!

(2) Suppose we have 38 consecutive numbers $N, N+1, \ldots, N+37$, none of which is deadly. By an analysis similar to that in part (1), none of the first nine can end in a 0. Hence, $N+9$ must end in a 0, as must $N+19$ and $N+29$. Then we must have $d(N+9) \equiv d(N+19) \equiv 1 \pmod{11}$. Therefore $d(N+18) \equiv 10 \pmod{11}$. Furthermore, if $N+18$ ends in k 9's we must have $k \equiv 6 \pmod{11}$.

The smallest possible such number is 999999, yielding the 38 consecutive numbers 999981, 999982, ..., 1000018. Indeed, none of these numbers is deadly: their sums of digits are congruent to 1, 2, ..., 10, 1, 2, ..., 10, 1, 2, ..., 10, 2, 3, ..., 9, and 10 (mod 11), respectively.

21. [APMO 1998] Find the largest integer n such that n is divisible by all positive integers less than $\sqrt[3]{n}$.

Solution: The answer is 420, which satisfies the condition since $7 < \sqrt[3]{420} < 8$ and $420 = \operatorname{lcm}\{1, 2, 3, 4, 5, 6, 7\}$.

Suppose $n > 420$ is an integer such that every positive integer less than $\sqrt[3]{n}$ divides n. Then $\sqrt[3]{n} > 7$, so $420 = \operatorname{lcm}(1, 2, 3, 4, 5, 6, 7)$ divides n; thus $n \geq 840$ and $\sqrt[3]{n} > 0$. Thus $2520 = \operatorname{lcm}(1, 2, \ldots, 9)$ divides n and $\sqrt[3]{n} > 13$. Now let m be the largest positive integer less than $\sqrt[3]{n}$; that is, $m < \sqrt[3]{n} \leq m+1$. We have $m \geq 13$ and $\operatorname{lcm}(1, 2, \ldots, m)$ divides n. But

$$\operatorname{lcm}(m-3, m-2, m-1, m) \geq \frac{m(m-1)(m-2)(m-3)}{6}, \qquad (\dagger)$$

since 2 and 3 are the only possible common divisors of these four numbers. Thus

$$\frac{m(m-1)(m-2)(m-3)}{6} \leq n \leq (m+1)^3,$$

implying that

$$m \leq 6\left(1+\frac{2}{m-1}\right)\left(1+\frac{3}{m-2}\right)\left(1+\frac{4}{m-3}\right).$$

The left-hand side of the inequality is an increasing function of m, and the right-hand side is a decreasing function of m. But for $m = 13$, we have

$$13 \cdot 12 \cdot 11 \cdot 10 = 17160 > 16464 = 6 \cdot 14^3,$$

so this inequality is false for all $m \geq 13$. Thus no $n > 420$ satisfies the given condition.

Note: Ryan Ko pointed out that the inequality (†) can be improved to

$$\operatorname{lcm}(m-3, m-2, m-1, m) \geq \frac{(m-1)(m-2)(m-3)(m-4)}{2}.$$

Why?

22. [USAMO 1991] Show that for any fixed positive integer n, the sequence

$$2,\ 2^2,\ 2^{2^2},\ 2^{2^{2^2}},\ \ldots \pmod{n}$$

is eventually constant. (The tower of exponents is defined by $a_1 = 2$ and $a_{i+1} = 2^{a_i}$ for every positive integer i.)

Proof: We apply strong induction on n. The base case $n = 1$ is clearly true. Assume that the conclusion is true for $n \leq k$, where k is some positive integer. We consider the case $n = k + 1$.

If $n = k + 1$ is odd, $2^{\varphi(n)} \equiv 1 \pmod{n}$ by Euler's theorem. Because $\varphi(n) < n$, by the induction hypothesis, the sequence $a_1, a_2, \ldots$ is eventually constant modulo $\varphi(n)$; that is, $a_i \equiv c \pmod{\varphi(n)}$ for large i. Consequently,

$$a_{i+1} \equiv 2^{a_i} \equiv 2^c \pmod{n}$$

is constant, completing the inductive step for this case.

If $n = k+1$ is even, we write $n = k+1 = 2^q \cdot m$ for some positive integer k and odd positive integer m. By the induction hypothesis, the sequence $a_1, a_2, \ldots$ is eventually constant modulo m. Clearly, $a_i \equiv 0 \pmod{2^q}$ for all sufficiently large i. Because 2^q and m are relatively prime, each of 2^q and m divides $a_{i+1} - a_i$, which implies that $n = 2^q \cdot m$ divides $a_{i+1} - a_i$; that is, the sequence $a_1, a_2, \ldots$ is eventually constant modulo $n = k + 1$, completing the inductive step for this case, and our induction is complete.

23. Prove that for $n \geq 5$, $f_n + f_{n-1} - 1$ has at least $n + 1$ prime factors.

Proof: For each $k \geq 1$, we have

$$f_{k+1} + f_k - 1 = 2^{2^{k+1}} + 2^{2^k} + 1 = (2^{2^k} + 1)^2 - (2^{2^{k-1}})^2$$
$$= (2^{2^k} + 1 - 2^{2^{k-1}})(2^{2^k} + 1 + 2^{2^{k-1}}).$$

Hence

$$f_{k+1} + f_k - 1 = a_k(f_k + f_{k-1} - 1), \qquad (*)$$

where $a_k = f_k - f_{k-1} + 1$.

We proceed by induction. We have

$$f_5 + f_4 - 1 = 3 \cdot 7 \cdot 13 \cdot 97 \cdot 241 \cdot 673,$$

and the property holds. Assume that for some $k \geq 5$, $f_k + f_{k-1} - 1$ has at least $k + 1$ prime factors. Using $(*)$ and the fact that

$$\gcd(f_k + f_{k-1} - 1, a_k) = \gcd(f_k + f_{k-1} - 1, f_k - f_{k-1} + 1)$$
$$= \gcd(f_k - f_{k-1} + 1, 2 \cdot 2^{2^{k-1}}) = 1,$$

we conclude that $f_{k+1} + f_k - 1$ has at least $k + 2$ prime factors, and we are done.

24. Prove that any integer can be written as the sum of the cubes of five integers, not necessarily distinct.

Proof: We use the identity $6k = (k + 1)^3 + (k - 1)^3 - k^3 - k^3$ for

$$k = \frac{n^3 - n}{6} = \frac{n(n-1)(n+1)}{6},$$

which is an integer for all n. We obtain

$$n^3 - n = \left(\frac{n^3 - n}{6} + 1\right)^3 + \left(\frac{n^3 - n}{6} - 1\right)^3 - \left(\frac{n^3 - n}{6}\right)^3 - \left(\frac{n^3 - n}{6}\right)^3.$$

Therefore, n is equal to the sum

$$(-n)^3 + \left(\frac{n^3 - n}{6}\right)^3 + \left(\frac{n^3 - n}{6}\right)^3 + \left(\frac{n - n^3}{6} - 1\right)^3 + \left(\frac{n - n^3}{6} + 1\right)^3.$$

Remark: One can prove that any rational number is the sum of the cubes of three rational numbers.

25. Integer or fractional parts?

(1) [Czech and Slovak 1998] Find all real numbers x such that

$$x\lfloor x\lfloor x\lfloor x\rfloor\rfloor\rfloor = 88.$$

(2) [Belarus 1999] Show that the equation

$$\{x^3\} + \{y^3\} = \{z^3\}$$

has infinitely many rational noninteger solutions.

Solution:

(1) Let $f(x) = x\lfloor x\lfloor x\lfloor x\rfloor\rfloor\rfloor$.

We claim that if a and b are real numbers with the same sign and $|a| > |b| \geq 1$, then $|f(a)| > |f(b)|$. We notice that $|\lfloor a\rfloor| \geq |\lfloor b\rfloor| \geq 1$. Multiplying this by $|a| > |b| \geq 1$, we have $|a\lfloor a\rfloor| > |b\lfloor b\rfloor| \geq 1$. Notice that $a\lfloor a\rfloor$ and $a\lfloor a\lfloor a\rfloor\rfloor$ have the same signs as $b\lfloor b\rfloor$ and $b\lfloor b\lfloor b\rfloor\rfloor$ respectively. In a similar manner,

$$|a\lfloor a\lfloor a\rfloor\rfloor| > |b\lfloor b\lfloor b\rfloor\rfloor| \geq 1, \quad |\lfloor a\lfloor a\lfloor a\rfloor\rfloor\rfloor| \geq |\lfloor b\lfloor b\lfloor b\rfloor\rfloor\rfloor| \geq 1,$$

and $|f(a)| > |f(b)|$, establishing our claim.

We have $f(x) = 0$ for $|x| < 1$, $f(1) = f(-1) = 1$. Suppose that $f(x) = 88$. So $|x| > 1$, and we consider the following two cases.

In the first case,we assume that $x \geq 1$. It is easy to check that $f\left(\frac{22}{7}\right) = 88$. From our claim, we know that $f(x)$ is increasing for $x > 1$. So $x = \frac{22}{7}$ is the unique solution on this interval.

In the second case, we assume that $x \leq -1$. From our claim, we know that $f(x)$ is decreasing for $x < 1$. Since

$$|f(-3)| = 81 < f(x) = 88 < \left|f\left(\frac{112}{37}\right)\right| = 112,$$

$-3 > x > -\frac{112}{37}$ and $\lfloor x\lfloor x\lfloor x\rfloor\rfloor\rfloor = -37$. But then $x = -\frac{88}{37} > -3$, a contradiction. Thus there is no solution on this interval.

Therefore, $x = \frac{22}{7}$ is the only solution.

Finally, we note that $\frac{22}{7}$ and $-\frac{112}{37}$ are found by finding $\lfloor x\rfloor$, $\lfloor x\lfloor x\rfloor\rfloor$, and $\lfloor x\lfloor x\lfloor x\rfloor\rfloor\rfloor$ in that order. For example, for $x \geq 1$, $f(3) < 88 < f(4)$, and so $3 < x < 4$. Then $\lfloor x\rfloor = 3$ and $x\lfloor x\lfloor 3x\rfloor\rfloor = 88$. Then $f(3) < 88 < f(10/3)$, so $\lfloor x\lfloor x\rfloor\rfloor = 9$, and so on.

(2) Let

$$x = \frac{3}{5} \cdot (125k + 1), \ y = \frac{4}{5} \cdot (125k + 1), \text{ and } z = \frac{6}{5} \cdot (125k + 1)$$

for every integer k. These are never integers because 5 does not divide $125k + 1$. Moreover, we note that

$$125x^3 = 3^3(125k + 1)^3 \equiv 3^3 \pmod{125}.$$

Hence, 125 divides $125x^3 - 3^3$ and $x^3 - \left(\frac{3}{5}\right)^3$ is an integer. Thus,

$$\{x^3\} = \frac{27}{125}.$$

Similarly,

$$\{y^3\} = \frac{64}{125} \quad \text{and} \quad \{z^3\} = \frac{216}{125} - 1 = \frac{91}{125} = \frac{27}{125} + \frac{64}{125},$$

implying that $\{x^3\} + \{y^3\} = \{z^3\}$.

26. Let n be a given positive integer greater than 1. If p is a prime divisor of the Fermat number f_n, prove that $p - 1$ is divisible by 2^{n+2}.

Proof: Since $n > 1$, $f_{n-1} = 2^{2^{n-1}} + 1$ is defined. Note that

$$\begin{aligned}(f_{n-1})^{2^{n+1}} &= \left(2^{2^{n-1}} + 1\right)^{2^{n+1}} = \left(2^{2^n} + 1 + 2^{2^{n-1}+1}\right)^{2^n} \\ &= \left(f_n + 2^{2^{n-1}+1}\right)^{2^n}.\end{aligned}$$

By the binomial theorem, we obtain

$$\begin{aligned}(f_{n-1})^{2^{n+1}} &\equiv \left(f_n + 2^{2^{n-1}+1}\right)^{2^n} \equiv \left(2^{2^{n-1}+1}\right)^{2^n} \equiv \left(2^{2^n}\right)^{2^{n-1}+1} \\ &\equiv (f_n - 1)^{2^{n-1}+1} \equiv (-1)^{2^{n-1}+1} \equiv -1 \pmod{f_n},\end{aligned}$$

implying that f_n divides $(f_{n-1})^{2^{n+1}} + 1$. Since p divides f_n, p divides $(f_{n-1})^{2^{n+1}} + 1$, from which the desired conclusion follows by setting $a = f_{n-1}$ in introductory problem 1.49 (2).

27. [USAMO 1999 proposal, by Gerald Heuer] The sequence

$$\{a_n\}_{n=1}^{\infty} = \{1, 2, 4, 5, 7, 9, 10, 12, 14, 16, 17, \dots\}$$

of positive integers is formed by taking one odd integer, then two even integers, then three odd integers, etc. Express a_n in closed form.

Solution: The solution is similar to the second proof of Example 1.70. We claim that

$$a_n = 2n - \left\lfloor \frac{1+\sqrt{8n-7}}{2} \right\rfloor$$

for every positive integer n.

We rewrite the given sequence in blocks as

$$\{a_n\}_{n=1}^{\infty} = \{1; 2, 4; 5, 7, 9; 10, 12, 14, 16; 17, \dots\}.$$

Consider the sequence

$$\{b_n\}_{n=1}^{\infty} = \{1; 2, 2; 3, 3, 3; 4, 4, 4, 4; 5, \dots\}.$$

We show that

$$a_n + b_n = 2n \tag{$*$}$$

for all positive integers n. This is clear for $n = 1$ and $n = 2$. Within each block in each sequence, $a_{n+1} = a_n + 2$ and $b_{n+1} = b_n$, so if the relation $(*)$ holds for the first integer of a block, it holds for all integers in that block. If it is true for the last integer of a block, then it is true for the first integer of the next block because a_n and b_n each increase by 1. By induction, relation $(*)$ holds for every positive integer n.

It suffices to show that

$$b_n = \left\lfloor \frac{1+\sqrt{8n-7}}{2} \right\rfloor. \tag{$\dagger$}$$

If $b_n = k$, it is in the kth group and is preceded by at least $k-1$ groups containing $1 + 2 + \cdots + (k-1)$ terms. Considering also the fact that there are $n-1$ terms before b_n, we conclude that

$$1 + 2 + \cdots + (b_n - 1) \le n - 1,$$

or

$$\frac{b_n(b_n-1)}{2} \le n - 1.$$

Solving the above quadratic inequality for b_n gives

$$b_n \leq \frac{1+\sqrt{8n+7}}{2},$$

from which (†) follows, by noting that b_n is the largest integer satisfying this inequality.

28. [USAMO 1998, by Bjorn Poonen] Prove that for each $n \geq 2$, there is a set S of n integers such that $(a-b)^2$ divides ab for every distinct $a, b \in S$.

Proof: We will prove the assertion by induction on n that we can find such a set, all of whose elements are *nonnegative*. For $n = 2$, we may take $S = \{0, 1\}$.

Now suppose that for some $n \geq 2$, the desired set S_n of n nonnegative integers exists. Let L be the least common multiple of $(a-b)^2$ and ab, with (a, b) ranging over pairs of distinct elements from S_n. Define

$$S_{n+1} = \{L + a \; : \; a \in S\} \cup \{0\}.$$

Then S_{n+1} consists of $n+1$ nonnegative integers, since $L > 0$. If $\alpha, \beta \in S_{n+1}$ and either α of β is zero, then $(\alpha-\beta)^2$ divides $\alpha\beta$. If $L+a, L+b \in S_{n+1}$, with a, b distinct elements of S_n, then

$$(L+a)(L+b) \equiv ab \equiv 0 \pmod{(a-b)^2},$$

so $[(L+a)-(L+b)]^2$ divides $(L+a)(L+b)$, completing the inductive step.

29. [St. Petersburg 2001] Show that there exist infinitely many positive integers n such that the largest prime divisor of n^4+1 is greater than $2n$.

Proof: We claim first that there are infinitely many numbers that are prime divisors of m^4+1 for some m. Suppose to the contrary that there is only a finite number of such primes. Let $p_1, p_2, \ldots, p_k$ be all of them. Let p be any prime divisor of $(p_1p_2\cdots p_k)^4+1$. This number cannot equal any p_i. This contradicts our assumption, and establishes the claim.

Let $\mathcal{P}$ be the set of all numbers that are prime divisors of m^4+1 for some m. Pick any p from $\mathcal{P}$ and any integer m such that p divides m^4+1. Let r be the residue of m modulo p. It follows that $r < p$ and p divides both r^4+1 and $(p-r)^4+1$. Let n be the minimum of r and $p-r$. It follows that $n < \frac{p}{2}$ or $p > 2n$. If n can be obtained using the construction above, then it satisfies the desired condition. If it is constructed using the prime p, then p divides $n^4 + 1$. Thus, any such number n can be constructed with only a finite number of primes p. Since the set $\mathcal{P}$ is infinite, and for each integer m such a number n can be constructed, there is an infinite number of integers n satisfying the desired condition.

Note: The interested reader might want to solve the following more challenging problem which appeared in USAMO in 2006.

> For integral m, let $p(m)$ be the greatest prime divisor of m. By convention, we set $p(\pm 1) = 1$ and $p(0) = \infty$. Find all polynomials f with integer coefficients such that the sequence $\{p(f(n^2)) - 2n\}_{n\geq 0}$ is bounded above. (In particular, this requires $f(n^2) \neq 0$ for $n \geq 0$.)

30. [Hungary 2003] For a positive integer k, let $p(k)$ denote the greatest odd divisor of k. Prove that for every positive integer n,

$$\frac{2n}{3} < \frac{p(1)}{1} + \frac{p(2)}{2} + \cdots + \frac{p(n)}{n} < \frac{2(n+1)}{3}.$$

Proof: Let

$$s(n) = \frac{p(1)}{1} + \frac{p(2)}{2} + \cdots + \frac{p(n)}{n}.$$

We need to show that

$$\frac{2n}{3} < s(n) < \frac{2(n+1)}{3}. \qquad (*)$$

We apply strong induction on n. The statement $(*)$ is true for $n = 1$ and $n = 2$, since

$$\frac{2 \cdot 1}{3} = \frac{2}{3} < s(1) = 1 < \frac{2(1+1)}{3} = \frac{4}{3}$$

and

$$\frac{2 \cdot 2}{3} = \frac{4}{3} < s(2) = 1 + \frac{1}{2} = \frac{3}{2} < \frac{2(2+1)}{3} = 2.$$

Assume that the statement $(*)$ is true for all integers n less than k, where k is some positive integer. We will show that the statement $(*)$ is true for integers $n = k + 1$. The key fact is that $p(2k) = p(k)$. We consider two cases.

In the first case, we assume that k is even. We write $k = 2m$, where m is a positive integer less than k. For $n = k + 1 = 2m + 1$, we note that

$$\begin{aligned} s(2m+1) &= \left(\frac{p(1)}{1} + \frac{p(3)}{3} + \cdots + \frac{p(2m+1)}{2m+1}\right) \\ &\quad + \left(\frac{p(2)}{2} + \frac{p(4)}{4} + \cdots + \frac{p(2m)}{2m}\right) \\ &= (m+1) + \left(\frac{p(1)}{2} + \frac{p(2)}{4} + \cdots + \frac{p(m)}{2m}\right) \\ &= (m+1) + \frac{1}{2}\left(\frac{p(1)}{1} + \frac{p(2)}{2} + \cdots + \frac{p(m)}{m}\right) \\ &= (m+1) + \frac{s(m)}{2}. \end{aligned}$$

By the induction hypothesis, we have

$$(m+1) + \frac{m}{3} < (m+1) + \frac{s(m)}{2} = s(2m+1) < (m+1) + \frac{(m+1)}{3}.$$

Since $\frac{2(2m+1)}{3} = \frac{4m+2}{3} < \frac{4m+3}{3} = (m+1) + \frac{m}{3}$ and $(m+1) + \frac{(m+1)}{3} = \frac{4(m+1)}{3} = \frac{2(2m+1+1)}{3}$, it follows that

$$\frac{2(2m+1)}{3} < s(2m+1) < \frac{2(2m+1+1)}{3},$$

which is $(*)$ for $n = 2m + 1$.

In the second case, we assume that k is odd. We write $k = 2m + 1$ and $n = k + 1 = 2m + 2$. Similar to the first case, we can show that

$$s(2m+2) = (m+1) + \frac{s(m+1)}{2}.$$

By the induction hypothesis, it is not difficult to show that the statement $(*)$ is also true for $n = 2m + 2$, which completes our induction.

31. If p^t is an odd prime power and m is an integer relatively prime to both p and $p - 1$, then for any a and b relatively prime to p,

$$a^m \equiv b^m \pmod{p^t} \text{ if and only if } a \equiv b \pmod{p^t}.$$

Proof: Since $(a - b)$ divides $(a^m - b^m)$, if p^t divides $(a - b)$ then p^t divides $(a^m - b^m)$.

Conversely, suppose a and b are relatively prime to p and $a^m \equiv b^m \pmod{p^t}$. Since m is relatively prime to both p and $p-1$, m is relatively prime to $(p-1)p^{t-1} = \varphi(p^t)$, so there exists a positive integer k such that $mk \equiv 1 \pmod{\varphi(p^t)}$. Then

$$a \equiv a^{mk} = (a^m)^k \equiv (b^m)^k = b^{mk} \equiv b \pmod{p^t},$$

as desired.

Note: We can view this as an additional property for Proposition 1.18. In Proposition 1.18 (f), if we have $a \equiv b \pmod{m}$, then for any positive integer k, $a^k \equiv b^k \pmod{m}$. This problem allows us to *take roots* for congruence relations under certain relations.

32. [Turkey 1997] Prove that for each prime $p \geq 7$, there exists a positive integer n and integers $x_1, \ldots, x_n, y_1, \ldots, y_n$ not divisible by p such that

$$\begin{aligned} x_1^2 + y_1^2 &\equiv x_2^2 \pmod{p}, \\ x_2^2 + y_2^2 &\equiv x_3^2 \pmod{p}, \\ &\vdots \\ x_n^2 + y_n^2 &\equiv x_1^2 \pmod{p}. \end{aligned}$$

Proof: We claim that $n = p-1$ satisfies the conditions of the problem. We first consider a system of equations

$$\begin{aligned} x_1^2 + y_1^2 &= x_2^2, \\ x_2^2 + y_2^2 &= x_3^2, \\ &\vdots \\ x_n^2 + y_n^2 &= x_{n+1}^2. \end{aligned}$$

We repeatedly use the most well-known Pythagorean triple $3^2 + 4^2 = 5^2$ to obtain the following equalities

$$\begin{aligned} (3^n)^2 + (3^{n-1} \cdot 4)^2 &= (3^{n-1} \cdot 5)^2, \\ (3^{n-1} \cdot 5)^2 + (3^{n-2} \cdot 5 \cdot 4)^2 &= (3^{n-2} \cdot 5^2)^2, \\ (3^{n-2} \cdot 5^2)^2 + (3^{n-3} \cdot 5^2 \cdot 4)^2 &= (3^{n-3} \cdot 5^3)^2, \\ &\cdots \\ (3^{n+1-i} \cdot 5^{i-1})^2 + (3^{n-i} \cdot 5^{i-1} \cdot 4)^2 &= (3^{n-i} \cdot 5^i)^2, \\ &\vdots \\ (3 \cdot 5^{n-1})^2 + (5^{n-1} \cdot 4)^2 &= (5^n)^2. \end{aligned}$$

Indeed, we set

$$x_i = 3^{n+1-i} \cdot 5^{i-1}, \quad y_i = 4 \cdot 3^{n-i} \cdot 5^{i-1},$$

for every $i = 1, \ldots, n$, and $x_{n+1} = 5^n$.

To finish our proof, we only need to note that by Fermat's little theorem, we have

$$x_{n+1}^2 - x_1^2 \equiv 5^{2n} - 3^{2n} \equiv 25^{p-1} - 9^{p-1} \equiv 0 \pmod{p}.$$

Note: There are infinitely many such n, for instance all multiples of $p-1$.

33. [HMMT 2004] For every positive integer n, prove that

$$\frac{\sigma(1)}{1} + \frac{\sigma(2)}{2} + \cdots + \frac{\sigma(n)}{n} \le 2n.$$

Proof: If d is a divisor of i, then so is $\frac{i}{d}$, and $\frac{i/d}{i} = \frac{1}{d}$. Summing over all divisors d of i (which is $\sigma(i)$), we see that $\frac{\sigma(i)}{i}$ is the sum of all the reciprocals of the divisors of i; that is,

$$\frac{\sigma(i)}{i} = \sum_{d|i} \frac{1}{d}$$

for every positive integer i. Consequently, the desired inequality becomes

$$\sum_{d|1} \frac{1}{d} + \sum_{d|2} \frac{1}{d} + \cdots + \sum_{d|n} \frac{1}{d} \le 2n.$$

As we have shown in the solution of introductory problem 27, if we write out all these summands on the left-hand side explicitly, each number $\frac{1}{d}$, with $1 \le d \le n$, appears $\left\lfloor \frac{n}{d} \right\rfloor$ times, once for each multiple of d that is less than or equal to n. Hence the desired inequality becomes

$$\frac{1}{1}\left\lfloor \frac{n}{1} \right\rfloor + \frac{1}{2}\left\lfloor \frac{n}{2} \right\rfloor + \frac{1}{3}\left\lfloor \frac{n}{3} \right\rfloor + \cdots + \frac{1}{n}\left\lfloor \frac{n}{n} \right\rfloor < 2n.$$

For each positive integer i, we have $\frac{1}{i}\left\lfloor \frac{n}{i} \right\rfloor < \frac{1}{i} \cdot \frac{n}{i} = \frac{n}{i^2}$. Hence it suffices to show that

$$\frac{n}{1^2} + \frac{n}{2^2} + \cdots + \frac{n}{n^2} < 2n,$$

or

$$\frac{1}{2^2} + \frac{1}{3^2} + \cdots + \frac{1}{n^2} < 1,$$

which follows from

$$\begin{aligned}\frac{1}{2^2} + \frac{1}{3^2} + \cdots + \frac{1}{n^2} &< \frac{1}{1 \cdot 2} + \frac{1}{2 \cdot 3} + \cdots + \frac{1}{n(n-1)} \\ &= \left(\frac{1}{1} - \frac{1}{2}\right) + \left(\frac{1}{2} - \frac{1}{3}\right) + \cdots + \left(\frac{1}{n-1} - \frac{1}{n}\right) \\ &= 1 - \frac{1}{n} < 1.\end{aligned}$$

Note: From calculus, we also know that

$$\frac{1}{1^2} + \frac{1}{2^2} + \cdots = \frac{\pi^2}{6} < 2.$$

34. [USAMO 2005, by Răzvan Gelca] Prove that the system

$$\begin{aligned}x^6 + x^3 + x^3y + y &= 147^{157}, \\ x^3 + x^3y + y^2 + y + z^9 &= 157^{147},\end{aligned}$$

has no solutions in integers x, y, and z.

First Proof: Add the two equations; then add 1 to each side to obtain

$$(x^3 + y + 1)^2 + z^9 = 147^{157} + 157^{147} + 1.$$

We prove that the two sides of this expression cannot be congruent modulo 19. We choose 19 because the least common multiple of the exponents 2 and 9 is 18, and by Fermat's little theorem, $a^{18} \equiv 1 \pmod{19}$ when a is not a multiple of 19. In particular, $(z^9)^2 \equiv 0$ or $1 \pmod{19}$, and it follows that the possible remainders when z^9 is divided by 19 are

$$-1,\ 0,\ 1.$$

Next calculate n^2 modulo 19 for $n = 0, 1, \ldots, 9$ to see that the possible residues modulo 19 are

$$-8,\ -3,\ -2,\ 0,\ 1,\ 4,\ 5,\ 6,\ 7,\ 9.$$

Consequently, adding a number from the last two lists gives the possible residues modulo 19 for $(x^3 + y + 1)^2 + z^9$:

	-8	-3	-2	0	1	4	5	6	7	9
-1	-9	-4	-3	-1	0	3	4	5	6	8
0	-8	-3	-2	0	1	4	5	6	7	9
1	-7	-2	-1	1	2	5	6	7	8	10

Finally, apply Fermat's little theorem to see that

$$147^{157} + 157^{147} + 1 \equiv 14 \pmod{19}.$$

Because we cannot obtain 14 (or -5), which does not appear in the table above, the system has no solution in integers x, y, and z.

Second Proof: We will show there is no solution to the system modulo 13. Add the two equations and add 1 to obtain

$$(x^3 + y + 1)^2 + z^9 = 147^{157} + 157^{147} + 1.$$

By Fermat's little theorem, $a^{12} \equiv 1 \pmod{13}$ when a is not a multiple of 13. Hence we compute $147^{157} \equiv 4^1 \equiv 4 \pmod{13}$ and $157^{147} \equiv 1^3 \equiv 1 \pmod{13}$. Thus

$$(x^3 + y + 1)^2 + z^9 \equiv 6 \pmod{13}.$$

The cubes modulo 13 are 0, ± 1, and ± 5. Writing the first given equation as

$$(x^3 + 1)(x^3 + y) \equiv 4 \pmod{13},$$

we see that there is no solution in the case $x^3 \equiv -1 \pmod{13}$ and for x^3 congruent to 0, 1, 5, -5 modulo 13. Correspondingly, $x^3 + y$ must be congruent to 4, 2, 5, -1. Hence

$$(x^3 + y + 1)^2 \equiv 12, 9, 10, \text{ or } 0 \pmod{13}.$$

Also, z^9 is a cube; hence z^9 must be 0, 1, 5, 8, or 12 modulo 13. The following table shows that 6 modulo 13 is not obtained by adding one of 0, 9, 10, 12 to one of 0, 1, 5, 8, 12:

	0	1	5	8	12
0	0	1	5	8	12
9	9	10	1	4	8
10	10	11	2	5	9
12	12	0	4	7	11

Hence the system has no solutions in integers.

Note: This argument shows that there is no solution even if z^9 is replaced by z^3.

35. [St. Petersburg 2000] What is the smallest number of weighings on a balance scale needed to identify the individual weights of a set of objects known to weigh $1, 3, 3^2, \dots, 3^{26}$ in some order? (The balance scale reports the weight of the objects in the left pan minus the weight of the objects in the right pan.)

Solution: At least three weighings are necessary: each of the first two weighings divides the weights into three categories (the weights in the left pan, the weights in the right pan, and the weights remaining off the scale). Because $27 > 3 \cdot 3$, some two weights must fall into the same category on both weighings, implying that these weights cannot be distinguished. We now show that three weighings indeed suffice.

Label the 27 weights using the three-letter words made up of the letters L, R, O. In the ith weighing, put the weights whose ith letter is L on the left pan and the weights whose ith letter is R on the right pan. The difference between the total weight of the objects in the left pan and the total weight of the objects in the right pan equals

$$\epsilon_0 3^0 + \epsilon_1 3^1 + \cdots + \epsilon_{26} 3^{26},$$

where ϵ_j equals $1, -1$, or 0 if 3^j is in the left pan, in the right pan, or off the scale, respectively. The value of the above sum uniquely determines all of the ϵ_j: the value of the sum modulo 3 determines ϵ_0, then the value of the sum modulo 9 determines ϵ_1; and so on.

Thus, for $j = 0, \dots, 26$, the ith weighing determines the ith letter of the weight that measures 3^j. After three weighings, we thus know exactly which weight measures 3^j, as desired.

Note: This is a case of a more general result, that each integer has a unique representation in base 3 using the digits $-1, 0, 1$. Clearly, this works for numbers n with $0 \le n < 3^1$ (since $0 = 0$, $1 = 1$, and $2 = 3 - 1$). Assume that this works for numbers n with $0 \le n < 3^k$ for some positive integer k. We consider n with $3^k \le n < 3^{k+1}$. If $3^k \le n < 2 \cdot 3^k$, it works because $n = 3^k + n_1$, where $0 \le n_1 < 3^k$; if $2 \cdot 3^k \le n < 3^{k+1}$, it works because $n = 3^{k+1} - 3^k + n_1$ with $0 \le n_1 < 3^k$. It is not difficult to see that it works for negative numbers and the representation is unique for every integer. Indeed, we can convert a regular base-3 representation easily to this new base-3 representation. For example,

$$\begin{aligned}49 = 1211_{(3)} &= 3^3 + 2 \cdot 3^2 + 3 + 1 \\ &= 2 \cdot 3^3 - 3^2 + 3 + 1 = 3^4 - 3^3 - 3^2 + 3 + 1.\end{aligned}$$

36. [Iberoamerican 1998] Let λ be the positive root of the equation $t^2 - 1998t - 1 = 0$. Define the sequence $x_0, x_1, \dots$ by setting

$$x_0 = 1, \quad x_{n+1} = \lfloor \lambda x_n \rfloor \qquad (n \geq 0).$$

Find the remainder when x_{1998} is divided by 1998.

Solution: We have

$$1998 < \lambda = \frac{1998 + \sqrt{1998^2 + 4}}{2} = 999 + \sqrt{999^2 + 1} < 1999,$$

$x_1 = 1998$, $x_2 = 1998^2$. Since $\lambda^2 - 1998\lambda - 1 = 0$,

$$\lambda = 1998 + \frac{1}{\lambda} \quad \text{and} \quad x\lambda = 1998x + \frac{x}{\lambda}$$

for all real numbers x. Since $x_n = \lfloor x_{n-1}\lambda \rfloor$ and x_{n-1} is an integer and λ is irrational, we have

$$x_n < x_{n-1}\lambda < x_n + 1, \quad \text{or} \quad \frac{x_n}{\lambda} < x_{n-1} < \frac{x_n + 1}{\lambda}.$$

Since $\lambda > 1998$, $\left\lfloor \frac{x_n}{\lambda} \right\rfloor = x_{n-1} - 1$. Therefore,

$$x_{n+1} = \lfloor x_n \lambda \rfloor = \left\lfloor 1998x_n + \frac{x_n}{\lambda} \right\rfloor = 1998x_n + x_{n-1} - 1,$$

that is, $x_{n+1} \equiv x_{n-1} - 1 \pmod{1998}$. Therefore by induction $x_{1998} \equiv x_0 - 999 \equiv 1000 \pmod{1998}$.

37. [USAMO 1996, by Richard Stong] Determine (with proof) whether there is a subset X of the integers with the following property: for any integer n there is exactly one solution of $a + 2b = n$ with $a, b \in X$.

First Proof: Yes, there is such a subset. As shown in introductory problem 39 (3), if the problem is restricted to the nonnegative integers, then the set of integers whose representations in base-4 contain only the digits 0 and 1 satisfies the desired property. To accommodate the negative integers as well, we switch to "base-(-4)." That is, we represent every integer in the form $\sum_{i=0}^{k} c_i(-4)^i$, with $c_i \in \{0, 1, 2, 3\}$ for all i and $c_k \neq 0$, and let X be the set of numbers whose representations use only the digits 0 and 1. This X will again have the desired property, once we show that every integer has a unique representation in this fashion.

To show that base-(-4) representations are unique, let $\{c_i\}$ and $\{d_i\}$ be two distinct finite sequences of elements of $\{0, 1, 2, 3\}$, and let j be the smallest integer such that $c_j \neq d_j$. Then

$$\sum_{i=0}^{k} c_i(-4)^i \not\equiv \sum_{i=0}^{k} d_i(-4)^i \pmod{4^j},$$

so in particular the two numbers represented by $\{c_i\}$ and $\{d_i\}$ are distinct. On the other hand, to show that n admits a base-(-4) representation, find an integer k such that

$$1 + 4^2 + \cdots + 4^{2k} \geq n$$

and express

$$n + 4 + \cdots + 4^{2k-1} = \sum_{i=0}^{2k} c_i 4^i.$$

Now set $d_{2i} = c_{2i}$ and $d_{2i-1} = 3 - c_{2i-1}$, and note that $n = \sum_{i=0}^{2k} d_i(-4)^i$.

Second Proof: For any S of integers, let $S^* = \{a + 2b \mid a, b \in S\}$. Call a finite set of integers $S = \{a_1, a_2, \ldots, a_m\}$ *good* if $|S^*| = |S|^2$; that is, if the values $a_i + 2a_j$ $(1 \leq i, j \leq m)$ are distinct. We first prove that given a good set and an integer n, we can always find a good superset T of S such that n is an element in T^*. If n is in S^* already, take $T = S$. Otherwise, take $T = S \cup \{k, n - 2k\}$, where k is to be chosen. Then put $T^* = S^* \cup Q \cup R$, where

$$Q = \{3k, 3(n - 2k), k + 2(n - 2k), (n - 2k) + 2k\}$$

and

$$R = \{k + 2a_i,\ (n - 2k) + 2a_i,\ a_i + 2k,\ a_i + 2(n - 2k) \mid 1 \leq i \leq m\}.$$

Note that for any choice of k, we have $n = (n - 2k) + 2k$ in Q, which is a subset of T^*. Except for n, the new values are distinct nonconstant linear forms in k, so if k is sufficiently large, they will all be distinct from each other and from the elements of S^*. This proves that T^* is good.

Starting with the good set $X_0 = \{0\}$, we thus obtain a sequence of sets $X_1, X_2, X_3, \ldots$ such that for each positive integer j, X_j is a good superset of X_{j-1} and X_j^* contains the jth term of the sequence $1, -1, 2, -2, 3, -3, \ldots$. It follows that

$$X = \bigcup_{j=0}^{\infty} X_j$$

has the desired property.

38. The number x_n is defined as the last digit in the decimal representation of the integer $\left\lfloor \sqrt{2}^n \right\rfloor$ $(n = 1, 2, \ldots)$. Determine whether the sequence $x_1, x_2, \ldots, x_n, \ldots$ is periodic.

Solution: The answer is negative.

Set $y_n = 0$ if x_n is even and $y_n = 1$ otherwise. The new sequence $y_1, y_2, \ldots, y_n, \ldots$ is formed by the residues of the numbers x_n modulo 2. If $x_1, x_2, \ldots, x_n, \ldots$ is periodic, then so is $y_1, y_2, \ldots, y_n, \ldots$. We shall prove that $y_1, y_2, \ldots, y_n, \ldots$ is not periodic, which implies that the answer to the question is negative.

Let us consider the sequence $y_1, y_3, y_5, \ldots, y_{2n+1}, \ldots$. Its term y_{2n+1} can be obtained as follows. Write down $\sqrt{2}$ in base-2, multiply by 2^n (this gives $(\sqrt{2})^{2n+1}$), and discard the fractional part of the result to get

$$\left\lfloor (\sqrt{2})^{2n+1} \right\rfloor .$$

Then take the last (binary) digit of this integer; it is y_{2n+1}. But multiplying by 2^n in base 2 amounts simply to shifting the binary point n positions to the right. This implies that y_{2n+1} is in fact the nth digit of $\sqrt{2}$ after the binary point. Since $\sqrt{2}$ is irrational, we conclude that the sequence $y_1, y_3, \ldots, y_{2n+1}, \ldots$ is not periodic. It is easy to infer from here that $y_1, y_2, \ldots, y_n, \ldots$ is not periodic too, and we are done.

39. [Erdös-Suranyi] Prove that every integer n can be represented in infinitely many ways as

$$n = \pm 1^2 \pm 2^2 \pm \cdots \pm k^2$$

for a convenient k and a suitable choice of the signs $+$ and $-$.

Proof: It suffices to prove the statement for nonnegative n's, because for negative n's we can simply change all the signs. The proof goes by induction of step 4; that is, establishing the statement for $n = k + 4$ based on the induction hypothesis for $n = k$.

We first show that the statement holds for $n = 0, 1, 2$, and 3. We consider representations 0, 1, 2 and 3:

$$0 = 1^2 + 2^2 - 3^2 + 4^2 - 5^2 - 6^2 + 7^2, \quad 1 = 1^2,$$
$$2 = -1^2 - 2^2 - 3^2 + 4^2, \quad 3 = -1^2 + 2^2.$$

If n is representable in the desired form then so is $n + 4$, because 4 can be written as

$$4 = (k+1)^2 - (k+2)^2 - (k+3)^2 + (k+4)^2 \qquad (*)$$

for any k. It follows inductively that a representation of the desired form can be written for any nonnegative integer n.

Note: From $(*)$ it also follows that

$$(k+1)^2 - (k+2)^2 - (k+3) + (k+4)^2$$
$$- (k+5)^2 + (k+6)^2 + (k+7)^2 - (k+8)^2 = 0$$

for every integer k; hence it can be easily inferred that the number of representations of an integer in the desired form is infinite.

40. [China 2004] Let n be a given integer with $n \geq 4$. For a positive integer m, let S_m denote the set $\{m, m+1, \ldots, m+n-1\}$. Determine the minimum value of $f(n)$ such that every $f(n)$-element subset of S_m (for every m) contains at least three pairwise relatively prime elements.

First Proof: The answer is

$$f(n) = \left\lfloor \frac{n+1}{2} \right\rfloor + \left\lfloor \frac{n+1}{3} \right\rfloor - \left\lfloor \frac{n+1}{6} \right\rfloor + 1. \qquad (*)$$

Let us call a set T *good* if T contains three (distinct) elements that are relatively prime.

In the first step, we establish two simple claims:

(a) $f(n)$ exists and $f(n) \leq n$;

(b) $f(n+1) \leq f(n) + 1$.

Since $n \geq 4$, $m, m+1, m+2, m+3$ are distinct element in S_m. If m is even, then the set $\{m+1, m+2, m+3\}$ is good; if m is odd, $\{m, m+1, m+2\}$ is good. Hence the n-element set S_m is good for all m, and so $f(n) \leq n$, establishing (a). Claim (b) follows directly from the relation

$$\{m, m+1, \ldots, m+n\} = \{m, m+1, \ldots, m+n-1\} \cup \{m+n\}.$$

Next we give a lower bound for $f(n)$. Consider $S_2 = \{2, 3, \ldots, n+1\}$ and its subset T_2 that contains those elements in S_2 that are multiples of either 2 or 3 or both. By the pigeonhole principle, any three elements in T must share a common factor (of either 2 or 3). Hence T_2 is not good. But by the inclusion and exclusion principle,

$$|T_2| = \left\lfloor \frac{n+1}{2} \right\rfloor + \left\lfloor \frac{n+1}{3} \right\rfloor - \left\lfloor \frac{n+1}{6} \right\rfloor,$$

and so

$$f(n) \geq \left\lfloor \frac{n+1}{2} \right\rfloor + \left\lfloor \frac{n+1}{3} \right\rfloor - \left\lfloor \frac{n+1}{6} \right\rfloor + 1, \qquad (**)$$

where $\lfloor x \rfloor$ is the greatest integer less than or equal to x. We claim that this lower bound is in fact the exact value of $f(n)$.

Since $n \geq 4$, we know that $m+1, m+2, m+3, m+4$ are distinct elements in S_m. If m is even, then $\{m+1, m+2, m+3\}$ is good; if m is odd, then $\{m+2, m+3, m+4)$ is good. Hence the n-element set S_m is good for all m. Using this fact with $(**)$ gives us $f(4) = 4$ and $f(5) = 5$.

By simple computation, the last inequality gives $f(4) \geq 4$, $f(5) \geq 5$, $f(6) \geq 5$, $f(7) \geq 6$, $f(8) \geq 7$, and $f(9) \geq 8$. Since $f(n) \leq n$, we conclude that $f(4) = 4$ and $f(5) = 5$. We claim that $f(6) = 5$. Then by claim (b), we have $f(7) = 6$, $f(8) = 7$, and $f(9) = 8$.

We now show that $f(6) = 5$; that is, any 5-element subset T of a set of 6 consecutive numbers is good. Among these 6 numbers, 3 are odd consecutive numbers (which is a good triple) and 3 are even consecutive numbers. If all three odd numbers are in T, then T is good and we are done. Otherwise, T must contain all the even numbers, and two of the three odd numbers. If the two odd numbers in T are consecutive (of the form $2x+1$ and $2x+3$), then T is good since $(2x+1, 2x+2, 2x+3)$ is in T; otherwise, the two odd numbers in T are of the form $2x+1$ and $2x+5$, and T is good since T contains both $(2x+1, 2x+2, 2x+5)$ and $(2x+1, 2x+4, 2x+5)$, and at least one of these two triples is good (since at least one of $2x+1$ and $2x+5$ is not divisible by 3).

Since it is clear that $f(n+1) \le f(n) + 1$, this combined with (**) gives $f(7) = 6$, $f(8) = 7$, and $f(9) = 8$.

Finally, we prove the result (*) by induction on n. The above arguments show that the base cases for $n \le 9$ are true. Assume that (*) is true for some $n = k$, where k is an integer greater than or equal to 9. For $n = k+1$, note that

$$\begin{aligned} S_m &= \{m, m+1, \ldots, m+k\} \\ &= \{m, m+1, \ldots, m+k-6\} \cup \{m+k-5, \ldots, m+k\}. \end{aligned}$$

Hence, by the pigeonhole principle, $f(k+1) \le f(k-5) + f(6) - 1$. Applying the induction hypothesis to $f(k-5)$, and using $f(6) = 5$, we have

$$\begin{aligned} f(k+1) &\le \left\lfloor \frac{k-4}{2} \right\rfloor + \left\lfloor \frac{k-4}{3} \right\rfloor - \left\lfloor \frac{k-4}{6} \right\rfloor + 5 \\ &= \left\lfloor \frac{k+2}{2} \right\rfloor + \left\lfloor \frac{k+2}{3} \right\rfloor - \left\lfloor \frac{k+2}{6} \right\rfloor + 1. \end{aligned}$$

This combined with (**) establishes (*) for $n = k+1$, and our induction is complete.

Second Proof: (By Kevin Modzelewski) We maintain the same notation as in the first solution. As we have shown in the first proof, all 5-element subsets of a set of 6 consecutive integers are good. Now we consider some cases.

(i) In this case we assume that $n \equiv 0 \pmod 6$. We write $n = 6k$. We can partition the set S_m into k subsets of 6 consecutive integers. If $4k+1$ numbers are chosen, by the pigeonhole principle one these subsets contains 5 of the chosen numbers, and hence is good. On the other hand, each subset contains 4 numbers that are either divisible by 2 or 3 (those numbers that are congruent to 0, 2, 3, 4 modulo 6). The $4k$-element subset consisting of these numbers is not good. Hence $f(n) = 4k + 1 = 4\left\lfloor \frac{n}{6} \right\rfloor + 1$.

(ii) In this case we assume that $n \equiv 1 \pmod 6$. We write $n = 6k+1$. By (i) and observation (b) in the first solution, we have $f(n) = 4k+1$ or $f(n) = 4k+2$. On the other hand, there are $4k+1$ elements in $S_1 = \{2, 3, \ldots, n+1\} = \{2, 3, \ldots, 6k+2\}$ that are divisible by 2 or 3. Hence $f(n) = 4k+2 = 4\left\lfloor \frac{n}{6} \right\rfloor + 2$.

(iii) In this case we assume that $n \equiv 2 \pmod 6$. We write $n = 6k+2$. By (ii) and observation (b), we have $f(n) = 4k+2$ or $f(n) = 4k+3$. On the other hand, there are $4k+2$ elements in $S_1 = \{2, 3, \ldots, 6k+3\}$ that are divisible by 2 or 3. Hence $f(n) = 4k+3 = 4\left\lfloor \frac{n}{6} \right\rfloor + 3$.

(iv) In this case we assume that $n \equiv 3 \pmod 6$. We write $n = 6k+3$. Again, we have $f(n) = 4k+2$ or $f(n) = 4k+3$. On the other hand, there are $4k+2$ elements in $S_1 = \{2, 3, \ldots, 6k+4\}$ that are divisible by 2 or 3. Hence $f(n) = 4k+4 = 4\left\lfloor \frac{n}{6} \right\rfloor + 4$.

(v) In this case we assume that $n \equiv 4 \pmod 6$. We write $n = 6k+4$. We can partition set S_m into $\{6k+1, 6k+2, 6k+3, 6k+4\}$ and k subsets of 6 consecutive integers. Let T be a subset of S_m that is not good. Each of the 6-element subsets can have 4 elements in T. Also note that $6k+1$ and $6k+3$ cannot be both in T. Hence T can have at most $4k+3$ elements. Hence $f(n) \le 4k+4$. By (iv) and observation (b), we conclude that $f(n) = 4k+4 = 4\left\lfloor \frac{n}{6} \right\rfloor + 4$.

(vi) In this case we assume that $n \equiv 5 \pmod 6$. We write $n = 6k+5$. Again, we have $f(n) = 4k+4$ or $f(n) = 4k+5$. On the other hand, there are $4k+4$ elements in $S_1 = \{2, 3, \ldots, 6k+6\}$ that are divisible by 2 or 3. Hence $f(n) = 4k+5 = 4\left\lfloor \frac{n}{6} \right\rfloor + 5$.

Combining the above, we conclude that

$$f(n) = 4 \cdot \left\lfloor \frac{n}{6} \right\rfloor + \begin{cases} 1 & n \equiv 0 \pmod 6, \\ 2 & n \equiv 1 \pmod 6, \\ 3 & n \equiv 2 \pmod 6, \\ 4 & n \equiv 3 \pmod 6, \\ 4 & n \equiv 4 \pmod 6, \\ 5 & n \equiv 5 \pmod 6. \end{cases}$$

It is then not difficult to check that

$$f(n) = \left\lfloor \frac{n+1}{2} \right\rfloor + \left\lfloor \frac{n+1}{3} \right\rfloor - \left\lfloor \frac{n+1}{6} \right\rfloor + 1.$$

Note: Note also that $f(n)$ can be expressed as

$$f(n) = n - \left\lfloor \frac{n}{6} \right\rfloor - \left\lfloor \frac{n+2}{6} \right\rfloor + 1.$$

Indeed, the above expression might be more convenient in the second solution. The equivalence of the two expressions can be established by repeated applying the Hermite identity (Proposition 1.48) as follows:

$$n = \left\lfloor 2 \cdot \frac{n}{2} \right\rfloor = \left\lfloor \frac{n}{2} \right\rfloor + \left\lfloor \frac{n}{2} + \frac{1}{2} \right\rfloor = \left\lfloor \frac{n}{2} \right\rfloor + \left\lfloor \frac{n+1}{2} \right\rfloor,$$

$$\left\lfloor \frac{n}{2} \right\rfloor = \left\lfloor 3 \cdot \frac{n}{6} \right\rfloor = \left\lfloor \frac{n}{6} \right\rfloor + \left\lfloor \frac{n+2}{6} \right\rfloor + \left\lfloor \frac{n+4}{6} \right\rfloor,$$

$$\left\lfloor \frac{n+1}{3} \right\rfloor = \left\lfloor 2 \cdot \frac{n+1}{6} \right\rfloor = \left\lfloor \frac{n+1}{6} \right\rfloor + \left\lfloor \frac{n+4}{6} \right\rfloor.$$

41. [China 1999] Find the least positive integer r such that for any positive integers a, b, c, d, $((abcd)!)^r$ is divisible by the product of

$$\begin{array}{llll} (a!)^{bcd+1}, & (b!)^{acd+1}, & (c!)^{abd+1}, & (d!)^{abc+1}, \\ ((ab)!)^{cd+1}, & ((bc)!)^{ad+1}, & ((cd)!)^{ab+1}, & ((ac)!)^{bd+1}, \\ ((bd)!)^{ac+1}, & ((ad)!)^{bc+1}, & ((abc)!)^{d+1}, & ((abd)!)^{c+1}, \\ ((acd)!)^{b+1}, & ((bcd)!)^{a+1}. & & \end{array}$$

Solution: Let p denote the product of the 14 numbers. Setting $b = c = d = 1$, then $p = (a!)^{2+3\cdot 2+3\cdot 2} = (a!)^{14}$, implying that $r \geq 14$. We claim that $r = 14$. It suffices to show that p divides $((abcd)!)^{14}$.
We pair numbers $(a!)^{bcd+1}$ and $((bcd)!)^{a+1}$. Indeed, we have

$$(a!)^{bcd+1} \cdot ((bcd)!)^{a+1} = \left[(a!)^{bcd} \cdot (bcd)!\right]\left[((bcd)!)^a \cdot (a)!\right]$$

and its cyclic analogous forms. Likewise, we have

$$((ab!)^{cd+1} \cdot ((cd)!)^{ab+1} = \left[((ab)!)^{cd} \cdot (cd)!\right]\left[((cd)!)^{ab} \cdot (ab)!\right]$$

and its cyclic analogous forms. It is then not difficult to see that our claim follows Example 1.74 (1).

42. Two classics on L.C.M.

(1) Let $a_0 < a_1 < a_2 < \cdots < a_n$ be positive integers. Prove that

$$\frac{1}{\operatorname{lcm}(a_0, a_1)} + \frac{1}{\operatorname{lcm}(a_1, a_2)} + \cdots + \frac{1}{\operatorname{lcm}(a_{n-1}, a_n)} \leq 1 - \frac{1}{2^n}.$$

(2) Several positive integers are given not exceeding a fixed integer constant m. Prove that if every positive integer less than or equal to m is not divisible by any pair of the given numbers, then the sum of the reciprocals of these numbers is less than $\frac{3}{2}$.

Proof: While it is clear that (1) is a property of L.C.M., it is not obvious that (2) is also related to L.C.M.

(1) We induct on n. The base case $n = 1$ is trivial, since $\text{lcm}(a_0, a_1) \ge \text{lcm}(1, 2) = 2$. We assume that the statement is true for $n = k$; that is, if $a_0 < a_1 < a_2 < \cdots < a_k$ are positive integers, then

$$\frac{1}{\text{lcm}(a_0, a_1)} + \frac{1}{\text{lcm}(a_1, a_2)} + \cdots + \frac{1}{\text{lcm}(a_{k-1}, a_k)} \le 1 - \frac{1}{2^k}.$$

Now we consider the case $n = k + 1$. Let $a_0 < a_1 < a_2 < \cdots < a_k < a_{k+1}$ be positive integers. We consider two cases.

- In the first case, we assume that $a_{k+1} \ge 2^{k+1}$. Then we have $\text{lcm}(a_k, a_{k+1}) \ge a_{k+1} \ge 2^{k+1}$. It follows by the induction hypothesis that

$$\frac{1}{\text{lcm}(a_0, a_1)} + \cdots + \frac{1}{\text{lcm}(a_{k-1}, a_k)} + \frac{1}{\text{lcm}(a_k, a_{k+1})} \le 1 - \frac{1}{2^k} + \frac{1}{2^{k+1}} = 1 - \frac{1}{2^{k+1}},$$

establishing the inductive step.

- In the second case, we assume that $a_k < 2^{k+1}$. We have

$$\frac{1}{\text{lcm}(a_{i-1}, a_i)} = \frac{\gcd(a_{i-1}, a_i)}{a_{i-1}a_i} \le \frac{a_1 - a_{i-1}}{a_{i-1}a_1} = \frac{1}{a_{i-1}} - \frac{1}{a_i}.$$

Adding the above inequalities for i from 1 through $k + 1$ gives

$$\frac{1}{\text{lcm}(a_0, a_1)} + \cdots + \frac{1}{\text{lcm}(a_{k-1}, a_k)} + \frac{1}{\text{lcm}(a_k, a_{k+1})} \le \frac{1}{a_0} - \frac{1}{a_{k+1}} \le 1 - \frac{1}{2^{k+1}},$$

again establishing the inductive step.

(2) The key is to interpret the sentence "every positive integer less than or equal to m is not divisible by any pair of the given numbers." Indeed, this implies that the least common multiple of every two of the given numbers is greater than m.

Given n numbers, denote them by $x_1, x_2, \ldots, x_n$. For a given i, there are $\left\lfloor \frac{m}{x_i} \right\rfloor$ multiples of x_i among $1, 2, \ldots, m$. None of them is a multiple of x_j for $j \ne i$, since the least common multiple of x_i and x_j is greater than m. Hence there are

$$\left\lfloor \frac{m}{x_1} \right\rfloor + \left\lfloor \frac{m}{x_2} \right\rfloor + \cdots + \left\lfloor \frac{m}{x_n} \right\rfloor$$

distinct elements in the set $\{1, 2, \ldots, m\}$ that are divisible by one of the numbers $x_1, x_2, \ldots, x_n$. None of these elements can be 1 (unless $n = 1$, in which case the claim is obvious). Hence

$$\left\lfloor \frac{m}{x_1} \right\rfloor + \left\lfloor \frac{m}{x_2} \right\rfloor + \cdots + \left\lfloor \frac{m}{x_n} \right\rfloor \le m - 1.$$

Taking into account that $\frac{m}{x_i} < \left\lfloor \frac{m}{x_i} \right\rfloor + 1$ for each i, we obtain

$$m\left(\frac{1}{x_1} + \frac{1}{x_2} + \cdots + \frac{1}{x_n}\right) < m + n - 1.$$

We now claim that $n \le \frac{m+1}{2}$, which will imply

$$\frac{1}{x_1} + \frac{1}{x_2} + \cdots + \frac{1}{x_n} < 1 + \frac{n-1}{m} < \frac{3}{2}.$$

Indeed, note that the greatest odd divisors of $x_1, x_2, \ldots, x_n$ are all distinct. Otherwise, if some two of the given numbers shared the same greatest odd divisor, one of them would be a multiple of the other, contradicting the hypothesis. Hence n does not exceed the number of odd integers among $1, 2, \ldots, m$, and our claim $n \le \frac{m+1}{2}$ follows.

43. For a positive integer n, let $r(n)$ denote the sum of the remainders of n divided by $1, 2, \ldots, n$. Prove that there are infinitely many n such that $r(n) = r(n-1)$.

Solution: By Proposition 1.46 (a), the remainder when n is divided by k is equal to $\left\{\frac{n}{k}\right\} \cdot k = n - \left\lfloor \frac{n}{k} \right\rfloor \cdot k$. Hence we have

$$r(n) = \sum_{k=1}^{n} \left(n - \left\lfloor \frac{n}{k} \right\rfloor \cdot k\right).$$

Thus, the condition $r(n) = r(n-1)$ is equivalent to the equation

$$\sum_{k=1}^{n} \left(n - \left\lfloor \frac{n}{k} \right\rfloor \cdot k\right) = \sum_{k=1}^{n-1} \left(n - 1 - \left\lfloor \frac{n-1}{k} \right\rfloor \cdot k\right),$$

or

$$2n - 1 = n + \sum_{k=1}^{n-1} [n - (n-1)] = \sum_{k=1}^{n} \left\lfloor \frac{n}{k} \right\rfloor \cdot k - \sum_{k=1}^{n-1} \left\lfloor \frac{n-1}{k} \right\rfloor \cdot k. \quad (*)$$

If k does not divide n, then $\lfloor \frac{n}{k} \rfloor = \lfloor \frac{n-1}{k} \rfloor$, and so $\lfloor \frac{n}{k} \rfloor \cdot k = \lfloor \frac{n-1}{k} \rfloor \cdot k$; if k divides n, then $\lfloor \frac{n}{k} \rfloor = \lfloor \frac{n-1}{k} \rfloor + 1$, and so $\lfloor \frac{n}{k} \rfloor \cdot k = \lfloor \frac{n-1}{k} \rfloor \cdot k + k$. We conclude that the equation $(*)$ is equivalent to

$$2n - 1 = \sum_{k|n} k.$$

But the last equation can easily be satisfied by setting $n = 2^m$, where m is a nonnegative integer. Indeed,

$$2n - 1 = 2^{m+1} - 1 = 1 + 2 + 2^2 + \cdots + 2^m.$$

Therefore, if n is a perfect power of 2, then $r(n) = r(n-1)$.

44. Two related IMO problems.

(1) [IMO 1994 Short List] A *wobbly number* is a positive integer whose digits are alternately nonzero and zero with the units digit being nonzero. Determine all positive integers that do not divide any wobbly numbers.

(2) [IMO 2004] A positive integer is called *alternating* if among any two consecutive digits in its decimal representation, one is even and the other is odd. Find all positive integers n such that n has a multiple that is alternating.

Solution: This is a continuation of introductory problem 52.

(1) If n is a multiple of 10, then the last digit of any of its multiples is 0, and so n does not divide any wobbly numbers. If n is a multiple of 25, then the last two digits of any of its multiples are 25, or 50, or 75, or 00, and so n does not divide any wobbly numbers. We now prove that these are the only numbers not dividing any wobbly numbers.

First, we consider odd numbers m not divisible by 5. Then $\gcd(m, 10) = 1$ and $\gcd((10^k - 1)m, 10) = 1$ for every positive integer k. By Euler's theorem, there exists an integer ℓ such that

$$10^\ell \equiv 1 \pmod{(10^k - 1)m},$$

implying that

$$10^{k\ell} \equiv 1 \pmod{(10^k - 1)m}.$$

Since

$$10^{k\ell} - 1 = \left(10^k - 1\right)\left(10^{k(\ell-1)} + 10^{k(\ell-2)} + \cdots + 10^k + 1\right),$$

we conclude that

$$w_k = \underbrace{\overline{101010\ldots1}}_{2\ell-1 \text{ digits}} = 10^{2(\ell-1)} + 10^{2(\ell-2)} + \cdots + 10^2 + 1$$

is divisible by m. In particular, w_2 is a wobbly number (with digits 0 and 1) divisible by m.

Second, we consider odd numbers m' that are divisible by 5. Since the number is not a multiple of 25, we can write $m' = 5m$. Then $5w_2$ is a wobbly number (with digits 0 and 5) divisible by m'.

Next, we consider perfect powers of 2. It suffices to show that 2^{2t+1} (for every nonnegative integer t) divides a $(2t-1)$-digit wobbly number. We induct on t. The base case $t = 1$ is trivial by considering wobbly numbers $v_1 = a_1 = 8$. For $t = 2$, we consider numbers in the form $v_2 = \overline{a_2 08} = 100a_2 + 8 = 4(25a_2 + 2)$. We need to find a nonzero digit a_2 such that $25a_2 + 2 \equiv 0 \pmod 8$. It is easy to check that $a_2 = 6$ satisfies the condition, and so 608 is a wobbly multiple of 2^5. In general, assume that 2^{2t+1} divides wobbly number $v_t = \overline{a_t 0 a_{t-1} \ldots 0 a_1}$. We write $v_t = 2^{2t+1} u_t$. Consider the numbers in the form

$$\overline{a_{t+1} 0 a_t 0 a_{t-1} \ldots 0 a_1} = a_{t+1} \cdot 10^{2t} + 2^{2t+1} u_t = 2^{2t}\left(5^{2t} a_{t+1} + 2u_t\right).$$

We need to find a digit a_{t+1} such that $5^{2t} a_{t+1} + 2u_t \equiv 8$. Since $S = \{0, 1, 2, 3, 4, 5, 6, 7, 8\}$ forms a complete set of residue classes modulo 8, there is an element a_{t+1} in S such that $5^{2t} a_{t+1} + 2u_t \equiv 8$, and for this a_{t+1}, the $(2t+1)$-digit wobbly number

$$v_{t+1} = \overline{a_{t+1} 0 a_t 0 a_{t-1} \ldots 0 a_1}$$

is divisible by 2^{2t+3}, completing the induction.

Finally, we consider the number of the form $2^t m$, where $t \geq 1$ and $\gcd(m, 10) = 1$. It suffices to show that $2^{2t+1} m$ divides a wobbly number. We claim that the concatenation of $\ell - 1$ $\overline{v_t 0} = v_t \cdot 10$'s and a v_t will do the job. Indeed,

$$\underbrace{\overline{v_t 0 v_t 0 \ldots v_t}}_{\ell\ v_t\text{'s}} = v_t \cdot w_{2t},$$

because 2^{2t+1} divides v_t and m divides w_{2t}.

(2) The answers are those positive integers that are not divisible by 20. We call an integer n an *alternator* if it has a multiple that is alternating. Because any multiple of 20 ends with an even digit followed by 0, multiples of 20 are not alternating. Hence multiples of 20 are not alternators. We show that all other numbers are alternators. Let n be a positive integer not a multiple of 20. Note that all divisors of an alternator are alternators. We may assume that n is a even number.

We first establish the following key fact:

> If $n = 2^\ell$ or $2 \cdot 5^\ell$, for some positive integer ℓ, then there exists a multiple $X(n)$ of n such that $X(n)$ is alternating and $X(n)$ has n digits.

Indeed, we can set

$$m = \frac{10^{n+1} - 10}{99} = \underbrace{\overline{101010\ldots10}}_{n \text{ digits}}.$$

For every integer $k = 0, 1, \ldots, n-1$, there exists a sequence $e_0, e_1, \ldots, e_k \in \{0, 2, 4, 6, 8\}$ such that

$$M + \sum_{i=0}^{k} e_i \cdot 10^i$$

is divisible by 2^{k+2} if n is of the form 2^ℓ, or by $2 \cdot 5^{k+1}$ if $n = 2 \cdot 5^\ell$. This is straightforwardly proved by induction on k (as we did in the proof of part (1) or Example 1.53). In particular, there exist $e_0, \ldots, e_{n-1} \in \{0, 2, 4, 6, 8\}$ such that

$$X(n) = m + \sum_{i=0}^{n-1} e_i \cdot 10^i$$

is divisible by n. This $X(n)$ is alternating and has n digits, establishing this fact.

Now we prove our main result. Because n is even and not divisible by 20, we write n in the form $n'm$, where $n' = 2^\ell$ or $2 \cdot 5^\ell$ and $\gcd(m, 10) = 1$. (Clearly, $n' \geq \ell$.) Let $c \geq n'$ be an integer such that $10^c \equiv 1 \pmod{m}$. (Such a c exists because $10^{\varphi(m)} \equiv 1 \pmod{m}$, by Euler's theorem.) Let M be the concatenation of $1010\ldots$ and $X'(n)$. More precisely, we set

$$M = \frac{10^{2mc+1} - 10}{99} \cdot 10^{n'} + X(n') = \overline{\underbrace{101010\ldots10}_{2mc \text{ digits}} X(n')}.$$

Since $X(n')$ is an alternating number with exactly n' digits, M is clearly an alternating number. Because $n' \geq \ell$, M is divisible by n'. Since $\gcd(2, m) = 1$, there exists $k \in \{0, 1, 2, \ldots, m-1\}$ such that $M \equiv -2k \pmod{m}$. We consider the number

$$X(n) = M + \sum_{i=1}^{k} 2 \cdot 10^{ci}.$$

Note that $10^c > X(n')$, because $c \geq n'$ and $X(n')$ has exactly n' digits (by the key fact we established earlier). It is not difficult to show that $X(n)$ is also alternating. It is clear that $X(n) \equiv m+2k \equiv 0 \pmod{m}$, that is, $X(n)$ is divisible by m. This $X(n)$ is also divisible by n' (since n' divides $10^{n'}$, which divides 10^c) and is alternating. Thus $X(n)$ is an alternating number divisible by n; that is, n is an alternator.

Note: There are different approaches to both parts. Nevertheless, all these methods work on powers of 2 and 5 first, and applying certain concatenations of wobbly/alternating numbers. These particular methods have been chosen because they are independent of the Chinese Remainder Theorem.

45. [USAMO 1995] Let p be an odd prime. The sequence $(a_n)_{n\geq 0}$ is defined as follows: $a_0 = 0$, $a_1 = 1, \ldots, a_{p-2} = p-2$, and for all $n \geq p-1$, a_n is the least positive integer that does not form an arithmetic sequence of length p with any of the preceding terms. Prove that for all n, a_n is the number obtained by writing n in base-$(p-1)$ and reading the result in base-p.

Proof: We say that a subset of positive integers is p-progression-free if it does not contain an arithmetic progression of length p. Denote by b_n the number obtained by writing n in base-$(p-1)$ and reading it in base-p. One can easily prove that $a_n = b_n$ for all $n = 0, 1, 2, \ldots$ by induction, using the following properties of the set $B = \{b_0, b_1, \ldots, b_n, \ldots\}$:

(a) B is p-progression-free;

(b) If $b_{n-1} < a < b_n$ for some $n \geq 1$, then the set $\{b_0, b_1, \ldots, b_{n-1}, a\}$ is not p-progression-free.

Indeed, assume that (a) and (b) hold. By the definitions of a_k and b_k, we have $a_k = b_k$ for $k = 0, 1, \ldots, p-2$. Let $a_k = b_k$ for all $k \leq n-1$, where $n \geq p-1$. By (a), the set

$$\{a_0, a_1, \ldots, a_{n-1}, b_n\} = \{b_0, b_1, \ldots, b_{n-1}, b_n\}$$

is p-progression-free, so $a_n \leq b_n$. Also, the inequality $a_n < b_n$ is impossible in view of (b). Hence $a_n = b_n$ and we are done.

It remains to establish properties (a) and (b). Let us note first that B consists of all numbers whose base-p representation does not contain the digit $p-1$. Hence (a) follows from the fact that if $a, a+d, \dots, a+(p-1)d$ is any arithmetic progression of length p, then all base-p digits occur in the base-p representation of its terms. To see this, represent d in the form $d = p^m k$, where $\gcd(k, p) = 1$. Then d ends in m zeros, and the digit δ preceding them is nonzero. It is easy to see that if α is the $(m+1)$st digit of a (from right to left), then the corresponding digits of $a, a+d, \dots, a+(p-1)d$ are the remainders of $\alpha, \alpha+\delta, \dots, \alpha+(p-1)\delta$ modulo p, respectively. It remains to note that $\alpha, \alpha+\delta, \dots, \alpha+(p-1)\delta$ is a complete set of residues modulo p, because δ is relatively prime to p. This finishes the proof of (a).

We start proving (b) by the remark that $b_{n-1} < a < b_n$ implies that a is not in B. Since B consists precisely of the numbers whose base-p representations do not contain the digit $p-1$, this very digit must occur in the base-p representation of a. Let d be the number obtained from a by replacing each of its digits by 0 if the digit is not $p-1$, and by 1 if it is $p-1$. Consider the progression

$$a-(p-1)d,\ a-(p-2)d,\ \dots,\ a-d,\ a.$$

As the definition of d implies, the first $p-1$ terms do not contain $p-1$ in their base-p representation. Hence, being less than a, they must belong to the set $\{b_0, b_1, \dots, b_{n-1}\}$. Therefore the set $\{b_0, b_1, \dots, b_{n-1}, a\}$ is not p-progression-free, and the proof is complete.

46. [IMO 2000] Determine whether there exists a positive integer n such that n is divisible by exactly 2000 different prime numbers, and 2^n+1 is divisible by n.

Solution: The answer is positive.

We claim the following key fact:

> For any integer $a > 2$ there exists a prime p such that p divides (a^3+1) but p does not divide $(a+1)$.

Indeed, since $a^3+1 = (a+1)(a^2-a+1)$, we need to show that there exists a prime p such that $p \mid (a^2-a+1)$ but $p \nmid (a+1)$. Since

$$a^2-a+1 = (a+1)(a-2)+3,$$

it follows that $\gcd(a^2-a+1, a+1) = 1$ or $\gcd(a^2-a+1, a+1) = 3$. In the first case, our claim is clearly true. In the second case, we note that 3 divides both $a+1$ and $a-2$, and so 3 fully divides a^2-a+1. Since $a>2$, $a^2-a+1>3$, and so there is a prime $p\neq 3$ that divides a^2-a+1, and this prime p satisfies the conditions of the claim.

By our claim, there exist (odd) distinct primes $p_1, p_2, p_3, \ldots, p_{2000}$ such that $p_1 = 3$, $p_2 \neq 3$, $p_2 \mid (2^{3^2}+1)$, and

$$p_{i+1} \mid (2^{3^{i+1}}+1), \quad p_{i+1} \nmid (2^{3^i}+1),$$

for every $2 \leq i \leq 1999$. It is not difficult to see that

$$n = p_1^{2000} \cdot p_2 \cdots p_{2000} = 3^{2000} p_2 \cdot p_{2000}$$

satisfies the conditions of the problem. Indeed, for every $2 \leq i \leq 2000$, $3^i \mid 3^{2000}$, and so

$$p_i \mid 2^{3^i}+1 \mid 2^{3^{2000}}+1.$$

By a simple induction, we also note that 3^{k+1} fully divides $2^{3^k}+1$ for every positive integer k, because 3 fully divides a^2-a+1 for $a = 2^{3^k}$ (as we have shown in the proof of our claim). Therefore,

$$n \mid 2^{3^{2000}}+1 \mid 2^n+1,$$

since n is a odd multiple of 3^{2000}.

47. Two cyclic symmetric divisibility relations.

(1) [Russia 2000] Determine whether there exist pairwise relatively prime integers a, b, and c with $a, b, c > 1$ such that

$$b \mid 2^a+1, \quad c \mid 2^b+1, \quad a \mid 2^c+1.$$

(2) [TST 2003, by Reid Barton] Find all ordered triples of primes (p, q, r) such that

$$p \mid q^r+1, \quad q \mid r^p+1, \quad r \mid p^q+1.$$

Solution: Order is the key word to this problem.

(1) The answer is negative. We claim that no such integers exist.

Assume for the sake of the contradiction that we did have pairwise relatively prime integers a, b, $c > 1$ such that b divides 2^a+1, c divides 2^b+1, and a divides 2^c+1. Then a, b, and c are all odd.

To make our life a bit easier, we first assume that all a, b, c are primes. By cyclic conditions given in the problem, we may assume that $a < b$ and $a < c$. By Fermat's little theorem and Proposition 1.30, $\operatorname{ord}_a(2) \mid \gcd(2c, a-1) = 2$, by noting that c is a prime greater than a. Since a is an odd prime, $\operatorname{ord}_a(2)$ must then be 2, implying that $a = 3$, and so $b \mid 2^a + 1 = 9$, which is a contradiction.

What if a, b, c are not all primes? We try to generalize our previous method. Let $\pi(n)$ denote the smallest prime factor of a positive integer n. We make the following claim:

> If p is a prime such that $p \mid (2^y + 1)$ and $p < \pi(y)$, then $p = 3$.

The proof of our claim is similar to our previous discussion for the case that all a, b, c are primes. Then $\operatorname{ord}_p(2) \mid \gcd(2y, p-1) = 2$. Again, we have $\operatorname{ord}_p(2) = 2$ and $p = 3$, establishing our claim.

Now we solve our main problem. Since a, b, c are pairwise relatively prime, $\pi(a)$, $\pi(b)$, and $\pi(c)$ are distinct. Without loss of generality, assume that $\pi(a) < \pi(b), \pi(c)$. Applying the claim with $(p, y) = (\pi(a), c)$, we find that $\pi(a) = 3$. Write $a = 3a_0$.

We claim that 3 fully divides a_0. Otherwise, 9 would divide $2^c + 1$ and hence $2^{2c} - 1$. Because $2^n \equiv 1 \pmod 9$ only if $6 \mid n$, we must have $6 \mid 2c$. Then $3 \mid c$, contradicting the assumption that a and c are relatively prime. Thus, 3 does not divide a_0, b, or c. Let $q = \pi(a_0bc)$, so that $\pi(q) = q \le \min\{\pi(b), \pi(c)\}$.

Suppose, for the sake of contradiction, that q divides a. Because a and c are relatively prime, q cannot divide c, implying that $\pi(q) = q$ is not equal to $\pi(c)$. Because $\pi(q) \le \pi(c)$, we must have $\pi(q) < \pi(c)$. Furthermore, q must divide $2^c + 1$ because it divides a factor of $2^c + 1$ (namely, a). Applying our claim with $(p, y) = (q, c)$, we find that $q = 3$, a contradiction. Hence, our assumption was wrong, and q does not divide a. Similarly, q does not divide c. It follows that q must divide b.

Now, let e be the order of 2 modulo q. Then $e \le q - 1$, so e has no prime factors less than q. Also, q divides b and hence $2^a + 1$ and $2^{2a} - 1$, implying that $e \mid 2a$. The only prime factors of $2a$ less than q are 2 and 3, so $e \mid 6$. Thus, $q \mid (2^6 - 1)$, and $q = 7$. However, $2^3 \equiv 1 \pmod 7$, so

$$2^a + 1 \equiv (2^3)^{a_0} + 1 \equiv 1^{a_0} + 1 \equiv 2 \pmod 7.$$

Hence, q does not divide $2^a + 1$, contradicting the assumption that q divides b, which divides $2^a + 1$.

(2) The answers are $(2, 5, 3)$ and cyclic permutations.

We check that this is a solution:

$$2 \mid 126 = 5^3 + 1, \quad 5 \mid 10 = 3^2 + 1, \quad 3 \mid 33 = 2^5 + 1.$$

Now let p, q, r be three primes satisfying the given divisibility relations. Since q does not divide $q^r + 1$, $p \neq q$, and similarly $q \neq r$, $r \neq p$, so p, q, and r are all distinct. We apply introductory problem 1.49 (1) in this solution.

We first consider the case that p, q, and r are all odd. Since $p \mid q^r + 1$, by introductory problem 49 (1), either $2r \mid p - 1$ or $p \mid q^2 - 1$. But $2r \mid p - 1$ is impossible because $2r \mid p - 1$ leads to $p \equiv 1 \pmod{r}$, or $0 \equiv p^q + 1 \equiv 2 \pmod{r}$, which contradicts the fact that $r > 2$. Thus we must have $p \mid q^2 - 1 = (q - 1)(q + 1)$. Since p is an odd prime and $q - 1$, $q + 1$ are both even, we must have $p \mid \frac{q-1}{2}$ or $p \mid \frac{q+1}{2}$; either way, $p \leq \frac{q+1}{2} < q$. But then by a similar argument we may conclude that $q < r$, $r < p$, a contradiction.

Thus, at least one of p, q, r must equal 2. By a cyclic permutation we may assume that $q = 2$. Now $p \mid 2^r + 1$, so by introductory problem 49 (1) again, either $2r \mid p - 1$ or $p \mid 2^2 - 1 = 3$. But $2r \mid p - 1$ is impossible as before, because r divides $p^q + 1 = p^2 + 1 = (p^2 - 1) + 2$ and $r > 2$. Hence, we must have $p = 3$, and $r \mid p^q + 1 = 3^2 + 1 = 10$. Because $r \neq q$, we must have $r = 5$. Hence $(2, 5, 3)$ and its cyclic permutations are the only solutions.

48. [IMO 2002 Short List] Let n be a positive integer, and let $p_1, p_2, \ldots, p_n$ be distinct primes greater than 3. Prove that $2^{p_1 p_2 \cdots p_n} + 1$ has at least 4^n divisors.

First Proof: We induct on n.

For $n = 1$, we consider the number $a_1 = 2^{p_1} + 1$. Since p_1 is odd, $2^{p_1} + 1 \equiv -1 + 1 \equiv 0 \pmod{3}$. Hence a_1 has distinct divisors 1, 3, and a_1 itself. Since $p_1 > 3$, it follows that $a_1 > 9$, and so $\frac{a_1}{3}$ is another divisor of a_1, implying that a_1 has at least 4 distinct divisors 1, 3, $\frac{a_1}{3}$, and a_1, establishing the base case.

Assume that the statement is true for $n = k$ for some positive integer k; that is, $a_k = 2^{p_1 p_2 \cdots p_k} + 1$ has at least 4^k divisors. We consider the case $n = k + 1$. Since $p_1, p_2, \ldots, p_{k+1}$ are odd, 3 divides both a_k and $2^{p_{k+1}} + 1$. Also, by introductory problem 38 (3),

$$\gcd(a_k, 2^{p_{k+1}} + 1) = \gcd(2^{p_1 p_2 \cdots p_k} + 1, 2^{p_{k+1}} + 1) = 3,$$

or

$$\gcd\left(a_k, \frac{2^{p_{k+1}}+1}{3}\right) = 1. \tag{$*$}$$

We note that both a_k and $2^{p_{k+1}}+1$ divide a_{k+1}, because $p_1 \cdots p_k$ and p_{k+1} are odd. We conclude that

$$a_{k+1} = a_k \cdot \frac{2^{p_{k+1}}+1}{3} \cdot b_k \tag{$**$}$$

for some integer b_k. By the induction hypothesis, and by ($*$), we conclude that the product

$$a_k \cdot \frac{2^{p_{k+1}}+1}{3}$$

has has at least $4^k \cdot 2$ divisors, namely, those 4^k divisors $d_1, d_2, \ldots, d_{4^k}$ of a_k and another 4^k divisors

$$d_i \cdot \frac{2^{p_{k+1}}+1}{3}$$

for every $1 \le i \le 4^k$. We arrange these $2 \cdot 4^k$ divisors in increasing order as $d_1 < d_2 < \cdots < d_{2 \cdot 4^k}$. By ($**$), these numbers are also divisors of a_{k+1}. We now consider numbers

$$d_1 b_k,\ d_2 b_k,\ \ldots,\ d_{2 \cdot 4^k} b_k.$$

They are also divisors of a_{k+1}. We claim that

$$d_1,\ d_2,\ \ldots,\ d_{2 \cdot 4^k},\ d_1 b_k,\ d_2 b_k,\ \ldots,\ d_{2 \cdot 4^k} b_k$$

are distinct divisors of a_{k+1}, from which our inductive step follows, since we find 4^{k+1} divisors of a_{k+1}. To establish our claim, it suffices to show that

$$d_1 b_k \ge d_{2 \cdot 4^k}.$$

Since $d_1 \ge 1$ and $d_{2 \cdot 4^k} \le a_k \cdot \frac{2^{p_{k+1}}+1}{3}$, it suffices to show that

$$b_k \ge a_k \cdot \frac{2^{p_{k+1}}+1}{3},$$

or

$$\left(a_k \cdot \frac{2^{p_{k+1}}+1}{3}\right)^2 \le a_{k+1},$$

by (∗∗). The last inequality is equivalent to

$$(2^{p_1p_2\cdots p_k}+1)^2(2^{p_{k+1}}+1)^2 \le 9(2^{p_1p_2\cdots p_{k+1}}+1),$$

which follows from the inequality

$$(2^u+1)^2(2^v+1)^2 \le 9(2^{uv}+1)$$

for integers u and v with u and v greater than or equal to 5. Indeed, we have

$$\begin{aligned}(2^u+1)^2(2^v+1)^2 &\le (2^{2u}+2\cdot 2^u+1)(2^{2v}+2\cdot 2^v+1)\\ &< (3\cdot 2^{2u}+1)(3\cdot 2^{2v}+1) < 9(2^{2u}+1)(2^{2v}+1)\\ &= 9(2^{2u+2v}+2^{2u}+2^{2v}+1) < 9(2^{2u+2v+2}+1)\\ &< 9(2^{uv}+1),\end{aligned}$$

since $uv-2u-2v-2=(u-2)(v-2)-6>3$.

Second Proof: (Based on work by Hyun Soo Kim) Call an integer "tenebrous" if it is odd, square-free, not divisible by 3, and at least 5. For any integer m, let $\psi(m)$ denote the number of distinct prime factors of m, and let $d(m)$ denote the number of factors of m. We wish to prove that $d(2^a+1)\ge 4^{\tau(a)}$ for all tenebrous integers a.

Induct on $\tau(a)$. For the base case $\tau(a)=1$, 2^a+1 is divisible by 3 exactly once and is greater than 3, so $\tau(2^a+1)\ge 2$ and $d(2^a+1)\ge 4$.

Now let a, b be relatively prime tenebrous integers such that the claim holds for both a and b. Clearly $2^{ab}+1$ is divisible by both 2^a+1 and 2^b+1, so we can write

$$2^{ab}+1 = C\cdot \operatorname{lcm}[2^a+1, 2^b+1].$$

Because $ab-2a-2b-4=(a-2)(b-2)-8>0$,

$$2^{ab}+1 > 2^{2a+2b+4} > (2^a+1)^2(2^b+1)^2 > \operatorname{lcm}[2^a+1,2^b+1]^2,$$

so $C\ge \operatorname{lcm}[2^a+1,2^b+1]$. From the comment, we have $\gcd(2^a+1,2^b+1)=3$, so as 3 divides each of 2^a+1 and 2^b+1 exactly once,

$$d(\operatorname{lcm}[2^a+1,2^b+1]) = \frac{d(2^a+1)d(2^b+1)}{2} \ge 2^{2\tau(a)+2\tau(b)-1}.$$

For every divisor m of $\operatorname{lcm}[2^a+1,2^b+1]$, both m and Cm are divisors of $2^{ab}+1$. Since $C>\operatorname{lcm}[2^a+1,2^b+1]$,

$$d(2^{ab}+1)\ge 2\cdot d(\operatorname{lcm}[2^a+1,2^b+1]) \ge 4^{\tau(a)+\tau(b)},$$

completing the induction.

Third Proof: (Based on work by Eric Price) Following the notation of the second solution, a stronger claim is that for any tenebrous integer a,

$$\tau(2^a + 1) \geq 2\tau(a).$$

We proceed by induction on $\tau(a)$. The base case is the same as in the first solution.

Now let a, b be coprime tenebrous integers. We claim that $\tau(2^{ab} + 1) \geq \tau(2^a + 1) + \tau(2^b + 1)$.

Note that

$$\frac{2^{ab}+1}{2^a+1} = \sum_{i=1}^{b} \binom{b}{i}(-2^a - 1)^{i-1} \equiv b - \binom{b}{2}(2^a + 1) \pmod{(2^a + 1)^2},$$

so if a prime p divides $2^a + 1$ exactly $k \geq 1$ times, then p divides $2^{ab} + 1$ either k times (if p doesn't divide b) or $k + 1$ times (if p divides b). In any case p divides $2^{ab} + 1$ at most twice as many times as p divides $2^a + 1$. The same is true for prime factors of $2^b + 1$.

As in the first solution, $2^{ab} + 1 > (2^a + 1)^2(2^b + 1)^2$, so in light of the above, $2^{ab} + 1$ must have a prime factor dividing neither $2^a + 1$ nor $2^b + 1$.

Clearly $2^{ab} + 1$ is divisible by $\text{lcm}[2^a + 1, 2^b + 1]$. Because $2^{ab} + 1$ has a prime factor not dividing $\text{lcm}[2^a + 1, 2^b + 1]$, we have

$$\begin{aligned}
\tau(2^{ab} + 1) &\geq \tau(\text{lcm}[2^a + 1, 2^b + 1]) + 1 \\
&= \tau(2^a + 1) + \tau(2^b + 1) - \tau(\gcd(2^a + 1, 2^b + 1)) + 1 \\
&= \tau(2^a + 1) + \tau(2^b + 1) - \tau(3) + 1 \\
&= \tau(2^a + 1) + \tau(2^b + 1),
\end{aligned}$$

completing the induction.

49. [Zhenfu Cao] Let p be a prime, and let $\{a_k\}_{k=0}^{\infty}$ be a sequence of integers such that $a_0 = 0$, $a_1 = 1$, and

$$a_{k+2} = 2a_{k+1} - pa_k$$

for $k = 0, 1, 2, \ldots$. Suppose that -1 appears in the sequence. Find all possible values of p.

Solution: The answer is $p = 5$. It is not difficult to see that it is a solution. For $p = 5$, $a_3 = -1$. Now we prove that it is the only solution.

Assume that $a_m = -1$ for some nonnegative integer m. Clearly, $p \neq 2$, because otherwise $a_{k+2} = 2a_{k+1} - 2a_k$ is even and -1 will not appear in the sequence. Thus, we can assume that $\gcd(2, p) = 1$. We consider the recursive relation

$$a_{k+2} = 2a_{k+1} - pa_k$$

modulo p, and then modulo $p - 1$. First, we obtain

$$a_{k+2} \equiv 2a_{k+1} \pmod{p},$$

implying that $a_{k+1} \equiv 2^k a_1 \mod p$. In particular, we have

$$-1 \equiv a_m \equiv 2^{m-1}a_1 \equiv 2^{m-1} \pmod{p}. \qquad (*)$$

Second, we obtain

$$a_{k+2} \equiv 2a_{k+1} - a_k \pmod{p-1},$$

or

$$a_{k+2} - a_{k+1} \equiv a_{k+1} - a_k \pmod{p-1};$$

that is, the sequence is arithmetic modulo $p - 1$. Hence

$$a_{k+1} \equiv (k+1)(a_1 - a_0) + a_0 \equiv k + 1 \pmod{p-1}.$$

In particular, we have

$$-1 \equiv a_m \equiv m \pmod{p-1},$$

or

$$m + 1 \equiv 0 \pmod{p-1}.$$

Since $\gcd(2, p) = 1$, by Fermat's little theorem, we have $2^{p-1} \equiv 1 \pmod{p}$. Combining the last two congruence relations and $(*)$, we have

$$1 \equiv 2^{m+1} \equiv 4 \cdot 2^{m-1} \equiv -4 \pmod{p},$$

implying that $5 \equiv 0 \pmod{p}$; that is, $p = 5$ is the only possible value.

50. [Qinsan Zhu] Let $\mathcal{F}$ be a set of subsets of the set $\{1, 2, \ldots, n\}$ such that

(a) if A is an element of $\mathcal{F}$, then A contains exactly three elements;

(b) if A and B are two distinct elements in $\mathcal{F}$, A and B share at most one common element;

Let $f(n)$ denote the maximum number of elements in $\mathcal{F}$. Prove that

$$\frac{(n-1)(n-2)}{6} \le f(n) \le \frac{(n-1)n}{6}.$$

Proof: We will begin with the upper-bound inequality, since it is easier to prove. For such a set $\mathcal{F}$, let us count the number of distinct doubletons $\{x, y\} \subset \{1, 2, \ldots, n\}$ that are subsets of some element of $\mathcal{F}$. Since any set $A \in \mathcal{F}$ contains 3 such distinct doubletons, and no two elements of $\mathcal{F}$ can share a common doubleton, it means that

$$3f(n) \le \binom{n}{2} = \frac{n(n-1)}{2},$$

so the right inequality is proved.

Now we prove the lower-bound inequality. The set $S = \{1, 2, \ldots, n\}$ has $\binom{n}{3} = \frac{n(n-1)(n-2)}{6}$ 3-element subsets. Let $\mathcal{T}$ denote the set of all these 3-element subsets. We consider the subsets

$$\mathcal{T}_i = \{\{a, b, c\} \mid \{a, b, c\} \in \mathcal{T}, \quad a + b + c \equiv i \pmod{n}\},$$

for $i = 0, 1, \ldots, n-1$. It is clear that these subsets are nonintersecting and their union is $\mathcal{T}$; that is, they form a partition of $\mathcal{T}$. Since $\mathcal{T}$ has $\binom{n}{3} = \frac{n(n-1)(n-2)}{6}$ elements, we may conclude by the pigeonhole principle that at least one of these n subsets has at least $\frac{n(n-1)(n-2)}{6n} = \frac{(n-1)(n-2)}{6}$ elements. Say $\mathcal{T}_j$ is such a subset. We claim that $\mathcal{T}_j$ satisfies both conditions (a) and (b).

It is clear $\mathcal{T}_j$ satisfies condition (a). For (b), assume (for contradiction) that there are two distinct elements A and B in $\mathcal{T}_j$ that share at least two elements. Assume that $A = \{x, y, z_1\}$ and $B = \{x, y, z_2\}$. Since A and B are elements of $\mathcal{T}_j$, we have $x + y + z_1 \equiv x + y + z_2 \equiv j \pmod{n}$, implying that $z_1 \equiv z_2 \pmod{n}$. But recall that $1 \le z_1, z_2 \le n$. So we must have $z_1 = z_2$, and thus $A = B$, a contradiction.

It follows that we can set $\mathcal{F} = \mathcal{T}_j$, and so $f(n)$ is at least the number of elements in $\mathcal{T}_j$; that is,

$$f(n) \ge \frac{(n-1)(n-2)}{6}.$$

Note: Under the same conditions, the last problem in the 6th Balkan Mathematics Olympiad (1989) was asking for

$$\frac{n(n-4)}{6} \le f(n) \le \frac{(n-1)n}{6}.$$

Qinsan Zhu improved this result when he encountered this problem during his preparation for the IMO 2004.

51. [IMO 1998] Determine all positive integers k such that

$$\frac{\tau(n^2)}{\tau(n)} = k,$$

for some n.

Note: Let $n = p_1^{a_1} p_2^{a_2} \cdots p_r^{a_r}$ be a prime decomposition of n. Then

$$\tau(n) = (a_1 + 1)(a_2 + 1) \cdots (a_r + 1)$$

and

$$\tau(n^2) = (2a_1 + 1)(2a_2 + 1) \cdots (2a_r + 1).$$

It follows that $\tau(n^2)$ is always odd, so if k is an integer, then it must be odd. We now prove that the converse is also true; that is, if k is an odd positive integer, then

$$k = \frac{\tau(n^2)}{\tau(n)} = \frac{(2a_1 + 1)(2a_2 + 1) \cdots (2a_r + 1)}{(a_1 + 1)(a_2 + 1) \cdots (a_r + 1)} \qquad (*)$$

for some nonnegative integers $a_1, a_2, \ldots, a_r$. (Since there are infinitely many primes, we can always set $n = p_1^{a_1} p_2^{a_2} \cdots p_r^{a_r}$.) We call a positive integer *acceptable* if can be written in the above form.

First Solution: A natural approach is strong induction on k. The result is trivial for $k = 1$ by setting $n = 1$, $r = 1$, and $a_1 = 0$.

For any odd integer $k > 1$, if it is of the form $4m + 1$, then

$$k = \frac{4m + 1}{2m + 1} \cdot 2m + 1.$$

Since $2m + 1 < k$, it is acceptable by the induction hypothesis. Hence k is also acceptable.

However, if k is of the form $4m+3$, then we further assume that it is of the form $8m+3$. Then we have

$$k = \frac{24m+9}{12m+5} \cdot \frac{12m+5}{6m+3} \cdot (2m+1),$$

and so k is acceptable by applying the induction hypothesis to $2m+1 < k$. Our proof remains open for the case $k = 8m+7$. We have to split into two more cases again. To terminate this process, we reformulate the above idea as follows.

Since every odd positive integer k can be written in the form $2^s x - 1$ for some positive integer x, it suffices to show that if x is acceptable, then so is $2^s x - 1$ for every $s \geq 1$. Let ℓ be such that

$$\frac{\tau(\ell^2)}{\tau(\ell)} = x.$$

If $s = 1$, then

$$k = 2^s x - 1 = 2x - 1 = \frac{2x-1}{x} \cdot x$$

shows that $k = 2x - 1$ is acceptable.

For $s > 1$, then

$$2^s x - 1 = \frac{2^s \cdot 3x - 3}{2^{s-1} \cdot 3x - 1} \cdot \frac{2^{s-1}3^2 x - 3}{2^{s-2}3^2 x - 1} \cdot \frac{2^{s-2}3^3 x - 3}{2^{s-3}3^3 x - 1} \cdots \frac{2^2 3^{k-2} x - 3}{2 \cdot 3^{k-2} x - 1} \cdot \frac{2 \cdot 3^{k-1} x - 3}{3^{k-1} x} \cdot x$$

shows that $k = 2^s x - 1$ is acceptable. Our induction is thus complete.

Second Proof: The proof is again by strong induction. Clearly the assertion is true for $k = 1$. Next assume that $k > 1$ is an odd positive integer and that the assertion is true for all positive odd integers less than k. As in the first solution, write $k = 2^s x - 1$, where x is an odd integer less than k. By the induction hypothesis, k_0 is acceptable.

It suffices to find $a_1, a_2, \ldots, a_r$ such that

$$k = x \cdot \frac{2a_1+1}{a_1+1} \cdot \frac{2a_2+1}{a_2+1} \cdots \frac{2a_t+1}{a_t+1}. \qquad (**)$$

Note that if we set $a_2 = 2a_1$, $a_3 = 2a_2$, and so on, the equation $(**)$ can be simplified to

$$2^s x - 1 = k = x \cdot \frac{2^t a_1 + 1}{a_1 + 1},$$

or

$$1 = 2^s x - \frac{2^t a_1 + 1}{a_1 + 1} \cdot x = \frac{2^s a_1 + 2^s - 2^t a_1 - 1}{a_1 + 1} \cdot x.$$

It is convenient to set $t = s$, and further reduce the above equation to

$$1 = \frac{2^s - 1}{a_1 + 1} \cdot x,$$

or $a_1 + 1 = (2^s - 1)x$. Combining the above, we conclude that equation (**) can be satisfied by setting $t = s$, $a_1 = (2^s - 1)x - 1$, $a_2 = 2a_1$, $a_3 = 2a_2, \ldots, a_t = 2a_{t-1}$. Our induction is complete.

52. [China 2005] Let n be a positive integer greater than two. Prove that the Fermat number f_n has a prime divisor greater than $2^{n+2}(n+1)$.

Proof: For $1 \le n \le 4$, we know that f_n are primes, and the conclusion is trivial. Now we assume that $n \ge 5$.

By introductory problem 49, we may assume that

$$f_n = p_1^{k_1} p_2^{k_2} \cdots p_m^{k_m}, \qquad (*)$$

where k is some positive integer, $p_1, \ldots, p_k$ are distinct primes, and $k_1, \ldots, k_m$ are positive integers with

$$p_i = 2^{n+1} x_i + 1$$

for some positive integer x_i, for every $1 \le i \le m$. It suffices to show that

$$x_i \ge 2(n+1) \qquad (**)$$

for some $1 \le i \le m$.

First, we give an upper bound for the sum $k_1 + k_2 + \cdots + k_m$. Note that for every i, $p_i \ge 2^{n+1} + 1$. It follows from $(*)$ and the binomial theorem that

$$2^{2^n} + 1 = f_n \ge (2^{n+1} + 1)^{k_1 + k_2 + \cdots + k_m} \ge 2^{(n+1)(k_1 + k_2 + \cdots + k_m)} + 1,$$

implying that

$$k_1 + k_2 + \cdots + k_m \le \frac{2^n}{n+1}. \qquad (\dagger)$$

Second, we give a lower bound for the sum $x_1 k_1 + x_2 k_2 + \cdots + x_m k_m$. By the binomial theorem again, we have

$$p_i^{k_i} \equiv (2^{n+1} x_i + 1)^{k_i} \equiv 2^{n+1} x_i k_i + 1 \pmod{2^{2n+2}}.$$

Since $2^n > 2n+2$ for $n \geq 5$, we have $f_n \equiv 1 \pmod{2^{2n+2}}$. Taking the equation $(*)$ modulo 2^{2n+2} gives

$$\begin{aligned} 1 &\equiv (2^{n+1}x_1k_1+1)(2^{n+1}x_2k_2+1)\cdots(2^{n+1}x_mk_m+1) \\ &\equiv 1+2^{n+1}x_1k_1+2^{n+1}x_2k_2+\cdots+2^{n+1}x_2k_2 \pmod{2^{2n+2}}, \end{aligned}$$

or

$$0 \equiv 2^{n+1}(x_1k_1+x_2k_2+\cdots+x_mk_m) \pmod{2^{2n+2}}.$$

It follows that

$$0 \equiv x_1k_1+x_2k_2+\cdots+x_mk_m \pmod{2^{n+1}}.$$

Since the x_i's and k_i's are nonnegative, we conclude that

$$x_1k_1+x_2k_2+\cdots+x_mk_m \geq 2^{n+1}. \tag{‡}$$

Let $x_i = \max\{x_1, x_2, \ldots, x_m\}$. Then inequality (‡) implies that

$$x_i(k_1+k_2+\cdots+k_m) \geq 2^{n+1}.$$

By inequality (†), we conclude that

$$x_i \geq \frac{2^{n+1}}{k_1+k_2+\cdots+k_m} \geq \frac{2^{n+1}}{\frac{2^n}{n+1}} = 2(n+1),$$

establishing the desired inequality $(**)$.

Glossary

Arithmetic function

A function defined on the positive integers that is complex valued.

Arithmetic-Geometric Means Inequality

If n is a positive integer and $a_1, a_2, \ldots, a_n$ are nonnegative real numbers, then

$$\frac{1}{n}\sum_{i=1}^{n} a_i \geq (a_1 a_2 \cdots a_n)^{1/n},$$

with equality if and only if $a_1 = a_2 = \cdots = a_n$. This inequality is a special case of the **power mean inequality**.

Base-b representation

Let b be an integer greater than 1. For any integer $n \geq 1$ there is a unique system $(k, a_0, a_1, \ldots, a_k)$ of integers such that $0 \leq a_i \leq b-1$, $i = 0, 1, \ldots, k$, $a_k \neq 0$ and

$$n = a_k b^k + a_{k-1} b^{k-1} + \cdots + a_1 b + a_0.$$

Beatty's theorem

Let α and β be two positive irrational real numbers such that

$$\frac{1}{\alpha} + \frac{1}{\beta} = 1.$$

The sets $\{\lfloor\alpha\rfloor, \lfloor 2\alpha\rfloor, \lfloor 3\alpha\rfloor, \ldots\}$, $\{\lfloor\beta\rfloor, \lfloor 2\beta\rfloor, \lfloor 3\beta\rfloor, \ldots\}$ form a partition of the set of positive integers.

Bernoulli's inequality

For $x > -1$ and $a > 1$,

$$(1+x)^a \geq 1 + ax,$$

with equality when $x = 0$.

Bézout's identity

For positive integers m and n, there exist integers x and y such that $mx + by = gcd(m, n)$.

Binomial coefficient

$$\binom{n}{k} = \frac{n!}{k!(n-k)!},$$

the coefficient of x^k in the expansion of $(x+1)^n$.

Binomial theorem

The expansion

$$(x+y)^n = \binom{n}{0}x^n + \binom{n}{1}x^{n-1}y + \binom{n}{2}x^{n-2}y + \cdots + \binom{n}{n-1}xy^{n-1} + \binom{n}{n}y^n.$$

Canonical factorization

Any integer $n > 1$ can be written uniquely in the form

$$n = p_1^{\alpha_1} \cdots p_k^{\alpha_k},$$

where $p_1, \ldots, p_k$ are distinct primes and $\alpha_1, \ldots, \alpha_k$ are positive integers.

Carmichael numbers

The composite integers n satisfying $a^n \equiv a \pmod{n}$ for every integer a.

Complete set of residue classes modulo n

A set S of integers such that for each $0 \leq i \leq n-1$ there is an element $s \in S$ with $i \equiv s \pmod{n}$.

Congruence relation

Let a, b, and m be integers, with $m \neq 0$. We say that a and b are congruent modulo m if $m \mid (a - b)$. We denote this by $a \equiv b \pmod{m}$. The relation "$\equiv$" on the set $\mathbb{Z}$ of integers is called the congruence relation.

Division algorithm

For any positive integers a and b there exists a unique pair (q, r) of nonnegative integers such that $b = aq + r$ and $r < a$.

Euclidean algorithm

Repeated application of the division algorithm:

$$\begin{aligned} m &= nq_1 + r_1, \ 1 \le r_1 < n, \\ n &= r_1 q_2 + r_2, \ 1 \le r_2 < r_1, \\ &\vdots \\ r_{k-2} &= r_{k-1} q_k + r_k, \ 1 \le r_k < r_{k-1}, \\ r_{k-1} &= r_k q_{k+1} + r_{k+1}, \ r_{k+1} = 0 \end{aligned}$$

This chain of equalities is finite because $n > r_1 > r_2 > \cdots > r_k$.

Euler's theorem

Let a and m be relatively prime positive integers. Then

$$a^{\varphi(m)} \equiv 1 \pmod{m}.$$

Euler's totient function

The function $\varphi(m)$ is defined to be the number of integers between 1 and n that are relatively prime to m.

Factorial base expansion

Every positive integer k has a unique expansion

$$k = 1! \cdot f_1 + 2! \cdot f_2 + 3! \cdot f_3 + \cdots + m! \cdot f_m,$$

where each f_i is an integer, $0 \le f_i \le i$, and $f_m > 0$.

Fermat's little theorem

Let a be a positive integer and let p be a prime. Then

$$a^p \equiv a \pmod{p}.$$

Fermat numbers

The integers $f_n = 2^{2^n} + 1, n \geq 0$.

Fibonacci sequence

The sequence defined by $F_0 = 1$, $F_1 = 1$, and $F_{n+1} = F_n + F_{n-1}$ for every positive integer n.

Floor function

For a real number x there is a unique integer n such that $n \leq x < n + 1$. We say that n is the greatest integer less than or equal to x or the floor of x and we write $n = \lfloor x \rfloor$.

Fractional part

The difference $x - \lfloor x \rfloor$ is called the fractional part of x and is denoted by $\{x\}$.

Fundamental theorem of arithmetic

Any integer n greater than 1 has a unique representation (up to a permutation) as a product of primes.

Hermite's identity

For any real number x and for any positive integer n,

$$\lfloor x \rfloor + \left\lfloor + \frac{1}{n} \right\rfloor + \left\lfloor + \frac{2}{n} \right\rfloor + \cdots + \left\lfloor + \frac{n-1}{n} \right\rfloor = \lfloor nx \rfloor.$$

Legendre's formula

For any prime p and any positive integer n,

$$e_p(n) = \sum_{i \geq 1} \left\lfloor \frac{n}{p^i} \right\rfloor.$$

Legendre's function

Let p be a prime. For any positive integer n, let $e_p(n)$ be the exponent of p in the prime factorization of $n!$.

Linear Diophantine equation

An equation of the form

$$a_1x_1 + \cdots + a_nx_n = b,$$

where $a_1, a_2, \ldots, a_n, b$ are fixed integers.

Mersenne numbers

The integers $M_n = 2^n - 1$, $n \geq 1$.

Möbius function

The arithmetic function μ defined by

$$\mu(n) = \begin{cases} 1 & \text{if } n = 1, \\ 0 & \text{if } p^2 \mid n \text{ for some prime } p > 1, \\ (-1)^k & \text{if } n = p_1 \cdots p_k, \text{ where } p_1, \ldots, p_k \text{ are distinct primes.} \end{cases}$$

Möbius inversion formula

Let f be an arithmetic function and let F be its summation function. Then

$$f(n) = \sum_{d|n} \mu(d) F\left(\frac{n}{d}\right).$$

Multiplicative function

An arithmetic function $f \neq 0$ with the property that for any relatively prime positive integers m and n,

$$f(mn) = f(m)f(n).$$

Number of divisors

For a positive integer n denote by $\tau(n)$ the number of its divisors. It is clear that

$$\tau(n) = \sum_{d|n} 1.$$

Order modulo m

We say that a has order d modulo m, denoted by $\text{ord}_m(a) = d$, if d is the smallest positive integer such that $a^d \equiv 1 \pmod{m}$.

Perfect number

An integer $n \geq 2$ with the property that the sum of its divisors is equal to $2n$.

Pigeonhole Principle

If n objects are distributed among $k < n$ boxes, some box contains at least two objects.

Prime number theorem

The relation

$$\lim_{n\to\infty} \frac{\pi(n)}{n/\log n} = 1,$$

where $\pi(n)$ denotes the number of primes less than or equal to n.

Prime number theorem for arithmetic progressions

For relatively prime integers a and r, let $\pi_{a,d}(n)$ denote the number of primes in the arithmetic progression $a, a+d, a+2d, a+3d, \ldots$ that are less than or equal to n. Then

$$\lim_{n\to\infty} \frac{\pi_{a,d}(n)}{n/\log n} = \frac{1}{\varphi(d)}.$$

This result was conjectured by Legendre and Dirichlet and proved by Charles De la Vallée Poussin.

Sum of divisors

For a positive integer n denote by $\sigma(n)$ the sum of its positive divisors including 1 and n itself. It is clear that

$$\sigma(n) = \sum_{d|n} d.$$

Summation function

For an arithmetic function f the function F defined by

$$F(n) = \sum_{d|n} f(d).$$

Wilson's theorem

For any prime p, $(p-1)! \equiv -1 \pmod{p}$.

Zeckendorf representation

Each nonnegative integer n can be written uniquely in the form

$$n = \sum_{k=0}^{\infty} \alpha_k F_k,$$

where $\alpha_k \in \{0, 1\}$ and $(\alpha_k, \alpha_{k+1}) \neq (1, 1)$ for each k.

Further Reading

1. Andreescu, T.; Feng, Z., *101 Problems in Algebra from the Training of the USA IMO Team*, Australian Mathematics Trust, 2001.

2. Andreescu, T.; Feng, Z., *102 Combinatorial Problems from the Training of the USA IMO Team*, Birkhäuser, 2002.

3. Andreescu, T.; Feng, Z., *103 Trigonometry Problems from the Training of the USA IMO Team*, Birkhäuser, 2004.

4. Andreescu, T.; Feng, Z., *A Path to Combinatorics for Undergraduate Students: Counting Strategies*, Birkhäuser, 2003.

5. Feng, Z.; Rousseau, C.; Wood, M., *USA and International Mathematical Olympiads 2005*, Mathematical Association of America, 2006.

6. Andreescu, T.; Feng, Z.; Loh, P., *USA and International Mathematical Olympiads 2004*, Mathematical Association of America, 2005.

7. Andreescu, T.; Feng, Z., *USA and International Mathematical Olympiads 2003*, Mathematical Association of America, 2004.

8. Andreescu, T.; Feng, Z., *USA and International Mathematical Olympiads 2002*, Mathematical Association of America, 2003.

9. Andreescu, T.; Feng, Z., *USA and International Mathematical Olympiads 2001*, Mathematical Association of America, 2002.

10. Andreescu, T.; Feng, Z., *USA and International Mathematical Olympiads 2000*, Mathematical Association of America, 2001.

11. Andreescu, T.; Feng, Z.; Lee, G.; Loh, P., *Mathematical Olympiads: Problems and Solutions from Around the World, 2001–2002*, Mathematical Association of America, 2004.

12. Andreescu, T.; Feng, Z.; Lee, G., *Mathematical Olympiads: Problems and Solutions from Around the World, 2000–2001*, Mathematical Association of America, 2003.

13. Andreescu, T.; Feng, Z., *Mathematical Olympiads: Problems and Solutions from Around the World, 1999–2000*, Mathematical Association of America, 2002.

14. Andreescu, T.; Feng, Z., *Mathematical Olympiads: Problems and Solutions from Around the World, 1998–1999*, Mathematical Association of America, 2000.

15. Andreescu, T.; Kedlaya, K., *Mathematical Contests 1997–1998: Olympiad Problems from Around the World, with Solutions*, American Mathematics Competitions, 1999.

16. Andreescu, T.; Kedlaya, K., *Mathematical Contests 1996–1997: Olympiad Problems from Around the World, with Solutions*, American Mathematics Competitions, 1998.

17. Andreescu, T.; Kedlaya, K.; Zeitz, P., *Mathematical Contests 1995–1996: Olympiad Problems from Around the World, with Solutions*, American Mathematics Competitions, 1997.

18. Andreescu, T.; Enescu, B., *Mathematical Olympiad Treasures*, Birkhäuser, 2003.

19. Andreescu, T.; Gelca, R., *Mathematical Olympiad Challenges*, Birkhäuser, 2000.

20. Andreescu, T., Andrica, D., *An Introduction to Diophantine Equations*, GIL Publishing House, 2002.

21. Andreescu, T.; Andrica, D., *360 Problems for Mathematical Contests*, GIL Publishing House, 2003.

22. Andreescu, T.; Andrica, D., *Complex Numbers from A to Z*, Birkhäuser, 2004.

23. Beckenbach, E. F.; Bellman, R., *An Introduction to Inequalities*, New Mathematical Library, Vol. 3, Mathematical Association of America, 1961.

24. Coxeter, H. S. M.; Greitzer, S. L., *Geometry Revisited*, New Mathematical Library, Vol. 19, Mathematical Association of America, 1967.

25. Coxeter, H. S. M., *Non-Euclidean Geometry*, The Mathematical Association of America, 1998.

26. Doob, M., *The Canadian Mathematical Olympiad 1969–1993*, University of Toronto Press, 1993.

27. Engel, A., *Problem-Solving Strategies*, Problem Books in Mathematics, Springer, 1998.

28. Fomin, D.; Kirichenko, A., *Leningrad Mathematical Olympiads 1987–1991*, MathPro Press, 1994.

29. Fomin, D.; Genkin, S.; Itenberg, I., *Mathematical Circles*, American Mathematical Society, 1996.

30. Graham, R.L.; Knuth, D.E.; Patashnik, O., *Concrete Mathematics*, Addison-Wesley, 1989.

31. Gillman, R., *A Friendly Mathematics Competition*, The Mathematical Association of America, 2003.

32. Greitzer, S.L., *International Mathematical Olympiads, 1959–1977*, New Mathematical Library, Vol. 27, Mathematical Association of America, 1978.

33. Holton, D., *Let's Solve Some Math Problems*, A Canadian Mathematics Competition Publication, 1993.

34. Kazarinoff, N.D., *Geometric Inequalities*, New Mathematical Library, Vol. 4, Random House, 1961.

35. Kedlaya, K; Poonen, B.; Vakil, R., *The William Lowell Putnam Mathematical Competition 1985–2000*, The Mathematical Association of America, 2002.

36. Klamkin, M., *International Mathematical Olympiads, 1978–1985*, New Mathematical Library, Vol. 31, Mathematical Association of America, 1986.

37. Klamkin, M., *USA Mathematical Olympiads, 1972–1986*, New Mathematical Library, Vol. 33, Mathematical Association of America, 1988.

38. Kürschák, J., *Hungarian Problem Book, volumes I & II*, New Mathematical Library, Vols. 11 & 12, Mathematical Association of America, 1967.

39. Kuczma, M., *144 Problems of the Austrian–Polish Mathematics Competition 1978–1993*, The Academic Distribution Center, 1994.

40. Kuczma, M., *International Mathematical Olympiads 1986–1999*, Mathematical Association of America, 2003.

41. Larson, L.C., *Problem-Solving Through Problems*, Springer-Verlag, 1983.

42. Lausch, H. *The Asian Pacific Mathematics Olympiad 1989–1993*, Australian Mathematics Trust, 1994.

43. Liu, A., *Chinese Mathematics Competitions and Olympiads 1981–1993*, Australian Mathematics Trust, 1998.

44. Liu, A., *Hungarian Problem Book III*, New Mathematical Library, Vol. 42, Mathematical Association of America, 2001.

45. Lozansky, E.; Rousseau, C. *Winning Solutions*, Springer, 1996.

46. Mitrinovic, D.S.; Pecaric, J.E.; Volonec, V. *Recent Advances in Geometric Inequalities*, Kluwer Academic Publisher, 1989.

47. Mordell, L.J., *Diophantine Equations*, Academic Press, London and New York, 1969.

48. Niven, I., Zuckerman, H.S., Montgomery, H.L., *An Introduction to the Theory of Numbers*, Fifth Edition, John Wiley & Sons, Inc., New York, Chichester, Brisbane, Toronto, Singapore, 1991.

49. Savchev, S.; Andreescu, T. *Mathematical Miniatures*, Anneli Lax New Mathematical Library, Vol. 43, Mathematical Association of America, 2002.

50. Sharygin, I.F., *Problems in Plane Geometry*, Mir, Moscow, 1988.

51. Sharygin, I.F., *Problems in Solid Geometry*, Mir, Moscow, 1986.

52. Shklarsky, D.O; Chentzov, N.N; Yaglom, I.M., *The USSR Olympiad Problem Book*, Freeman, 1962.

53. Slinko, A., *USSR Mathematical Olympiads 1989–1992*, Australian Mathematics Trust, 1997.

54. Szekely, G.J., *Contests in Higher Mathematics*, Springer-Verlag, 1996.

55. Tattersall, J.J., *Elementary Number Theory in Nine Chapters*, Cambridge University Press, 1999.

56. Taylor, P.J., *Tournament of Towns 1980–1984*, Australian Mathematics Trust, 1993.

57. Taylor, P.J., *Tournament of Towns 1984–1989*, Australian Mathematics Trust, 1992.

58. Taylor, P.J., *Tournament of Towns 1989–1993*, Australian Mathematics Trust, 1994.

59. Taylor, P.J.; Storozhev, A., *Tournament of Towns 1993–1997*, Australian Mathematics Trust, 1998.

60. Yaglom, I.M., *Geometric Transformations*, New Mathematical Library, Vol. 8, Random House, 1962.

61. Yaglom, I.M., *Geometric Transformations II*, New Mathematical Library, Vol. 21, Random House, 1968.

62. Yaglom, I.M., *Geometric Transformations III*, New Mathematical Library, Vol. 24, Random House, 1973.

Index

arithmetic functions, 36

base-b representation, 41
Beatty's theorem, 60
Bernoulli's inequality, 145
Bézout's identity, 13
binomial theorem, 5

canonical factorization, 8
Carmichael numbers, 32
ceiling, 52
Chinese remainder theorem, 22
complete set of residue classes, 24
composite, 5
congruence relation, 19
coprime, 11

decimal representation, 41
Diophantine equations, 14
division algorithm, 4

Euclidean algorithm, 12
Euler's theorem, 28
Euler's totient function, 27

factorial base expansion, 45
Fermat numbers, 22, 70
Fermat's little theorem, 28
Fibonacci
    numbers, 45
    sequence, 45
fractional part, 52
fully divides, 9

fundamental theorem of arithmetic, 7

geometric progression, 9
greatest common divisor, 11

Hermite identity, 63

inverse of a modulo m, 26

least common multiple, 16
    of $a_1, a_2, \ldots, a_n$, 16
Legendre function, 65
linear combinations, 14
linear congruence equation, 22
linear congruence system, 22
linear Diophantine equation, 38

Mersenne numbers, 71
multiplicative arithmetic functions, 18
Möbius function, 36
Möbius inversion formula, 37

number of divisors, 17

order d modulo m, 32

perfect cube, 2
perfect numbers, 72
perfect power, 2
perfect square, 2
pigeonhole principle, 93
prime, 5

prime number, 5

quotient, 4

reduced complete set of residue classes, 28
relatively prime, 11
remainder, 4

square free, 2
sum of positive divisors, 18
summation function, 36

twin primes, 6

Wilson's theorem, 26
Wolstenholme's theorem, 115

Zeckendorf representation, 45